FAST POLYMERIZATION PROCESSES

Polymer Science and Engineering Monographs: A State-of-the-Art Tutorial Series

A series edited by **Eli M. Pearce**, Polytechnic University, Brooklyn, New York

Associate Editors

Gennadi E. Zaikov, Russian Academy of Sciences, Moscow
Yasunori Nishijima, Kyoto University, Japan

Volume 1

Fast Polymerization Processes
Karl S. Minsker and *Alexandre Al. Berlin*

Additional volumes in preparation

Physical Properties of Polymers: Prediction and Control
A. A. Askadskii

Multicomponent Transport in Polymer Systems for Controlled Release
A. Ya. Polishchuk and *G. E. Zaikov*

This book is part of a series. The publisher will accept continuation orders which may be cancelled at any time and which provide for automatic billing and shipping of each title in the series upon publication. Please write for details.

FAST POLYMERIZATION PROCESSES

Karl S. Minsker
Bashkirian State University
Bashkortostan, Russian Federation

and

Alexandre Al. Berlin
Russian Academy of Sciences
Moscow

Gordon and Breach Publishers

Australia • China • France • Germany • India • Japan • Luxembourg • Malaysia
The Netherlands • Russia • Singapore • Switzerland • Thailand • United Kingdom
United States

Emmaplein 5
1075 AW Amsterdam
The Netherlands

British Library Cataloguing in Publication Data

Minsker, K. S.
Fast Polymerization Processes. –
(Polymer Science & Engineering
Monographs: A State-of-the-Art Tutorial Series,
ISSN 1023-7720; Vol. 1)
I. Title II. Berlin, A. III. Series
547.28

ISBN 2-88449-149-X (hardcover)
2-88449-191-0 (softcover)

CONTENTS

INTRODUCTION TO THE SERIES

This series will provide, in the form of single-topic volumes, state-of-the-art information in specific research areas of basic applied polymer science. Volumes may incorporate a brief history of the subject, its theoretical foundations, a thorough review of current practice and results, the relationship to allied areas, and a bibliography. Books in the series will act as authoritative references for the specialist, acquaint the non-specialist with the state of science in an allied area and the opportunity for application to his own work, and offer the student a convenient, accessible review that brings together diffuse information on a subject.

PREFACE

It is difficult to find a branch of chemical science that has never been studied. However, very fast chemical reactions in liquid phase, with reaction time lower than that of reagent mixing, represents a new field in chemistry and technology, both from the scientific and practical points of view. An example would be very fast isobutylene polymerization.

Essential new scientific propositions and methods have been discovered and introduced into practice, making it possible to consider fast chemical processes as an independent class of chemical reactions characterized by original regularities, specific control methods, technological fabrication of the process and the design of the main apparatus.

A great number of empirical methods of calculating chemical reactor designs exist. The main limitations of these methods are the introduction of many empirical parameters leading to difficulties in the scale conversion. However, methods of accurate calculation of the chemical reaction processes in various configurations—taking into account both chemical kinetics and mass- and heat transfer—recently have been developed. Such approaches were referred to by D. A. Frank-Kamenetsky as macrokinetics. This book gives examples of much analysis for fast chemical reaction occurring in reagent turbulent flows being mixed.

Due to the application of such approaches, we have managed to profoundly change the design of reactors for a number of fast chemical processes in "liquid-liquid" and "liquid-gas" systems such as isobutylene, a-olefin, piperilene oligomerization; ethylene, benzene and rubber chlorination; sulfuric acid alkylation; synthetic rubber production, and so on. The new reactors make it possible to considerably increase productivity, lower costs, power consumption and starting reagents, and to improve both product quality and ecological effects of the process.

This book will be of interest to researchers and engineers working in the chemical and petrochemical industries. It should prove useful for undergraduates specializing in chemical synthesis and technology.

The approaches and methods described in the book are still being developed and updated. Therefore, everything should not be taken as the "last" word. We hope that readers who become interested in these problems will watch the scientific papers appearing in this field of macrokinetics.

We are grateful to our late friend Academician Enikolopov/ Enikolopyan/ N.S., as well as to the colleagues, co-workers, and disciples whose names appear in the Reference section. Working with us for many years, they have devoted themselves to the fascinating and captivating search for new laws and new ways of solving many fundamental and technical problems.

We deeply appreciate the invaluable help of Farida Basirova, Svetiana Gantseva, and Sergei Minsker in preparing the manuscript.

INTRODUCTION

In chemistry and chemical engineering, especially the technology of the production of applied polymers, there are problems the solutions to which remain little studied or completely unknown. We are concerned with one of these: very fast polymerization in liquid-phase with the characteristic reaction times of the chemical reaction τ_{ch} less than 10^{-1}–10^{-3} s. A typical example of a similar process is isobutylene polymerization [1, 2].

An understanding of the fundamental laws and the effects of spontaneous heating of the reaction mass on the rate and the direction of fast chemical reactions as well as on the main mass-molecular characteristics of the final products (in polymer production-$\bar{M}_n$ and MWD) is of great significance. This is true both for controlling chemical processes in general and for selecting the design of the basic industrial equipment—polymerizers.

In order to obtain polymer products with the narrow, most probable inclusive MWD, to control the molecular characteristics, it is necessary to provide isothermal conditions of the reaction. This is usually achieved by removing the reaction heat evolved in exothermal processes and by the ideal mixing that provides a sufficiently uniform temperature field.

Theoretically, isothermal conditions in very fast reactions of polymerization can be created by using low concentrations of the monomer in the initial reaction mixture or by applying the catalyst for slower action. However, in these cases the reaction rate and the process productivity are considerably reduced. Besides, using great quantities of the solvent can seriously increase the power consumption and other expenses on the solvent regeneration. While the laboratory work can achieve well-known methods of obtaining isothermicity in polymerization, this is not true in the field. There, when working with very fast chemical processes, heat and mass transfer generally cannot be provided efficiently enough. Fast processes of liquid-phase polymerization are practically always characterized by high exothermicity of the reaction and low thermal conductivity of the reaction mixture. The result is local overheating of the reaction mass. Non-uniform time-spatial distributions of temperature, concentrations of the monomer and the catalyst and the degree of the reaction completion appear. As a result, very fast polymerization processes are hard to control and are characterized by a marked decrease of the total yield of the final product, poor quality, and unjustified retention of the reaction mixture in reactors-polymerizers. In a number of cases, temperature fluctuations of polymerization lead to unpredictable results (thermal explosion, powerful hydro- and pneumoshocks etc.).

Nevertheless, in realizing fast polymerization in industry, traditional methods were used to compute the designs of basic reactors-polymerizers and to develop the procedure generally. *A priori*, it is considered that fast reactions proceed under isothermal conditions. This is basically incorrect, since in very fast liquid-phase polymerization in volumetric reactors it is impossible to provide efficient heat and mass transfer in time commensurate with that of the chemical reaction.

Meanwhile, both in this country and abroad, the state of technology in realizing very fast polymerization processes in liquid-phase is such that polymers (of isobutylene, in particular) are obtained in volumetric reactors of ideal displacement, with continuous action with the volume of 2–30 m^3 at the intense mixing and heat removal.

The use of continuous action reactors of ideal displacement in liquid-phase fast polymerization is an even worse choice, as it results in still greater broadening of the MWD and the increase of the molecular mass of polymers formed in comparison to the results predicted by theoretical calculations for reactions proceeding under isothermal conditions. In practice, the isothermal process even for relatively slow chemical reactions cannot be realized in continuous action reactors of ideal displacement, since it is necessary to vary heat transfer along the reactor length according to the kinetics of heat evolution. The displacement reactors normally operate in adiabatic or intermediate, but far from isothermal, regimes (with outer heat removal excluded). Therefore, practically all reactors used in industry in fast chemical processes have essential shortcomings in design and, hence, as a rule, are inefficient in operation. Consequently, the quality of the products obtained is far from ideal, and processes in general are imperfect both in engineering and socio-economical aspects (that is, in terms of yield, quality, unrecovered waste, unjustified great power consumption, etc.).

Thus, realizing very fast chemical processes in real productions—in particular in that of liquid-phase electrophilic isobutylene polymerization—one should always bear in mind that the purely chemical aspect of the process is complicated by a number of physical processes starting with diffusion, mass and heat transfer, and hydrodynamic processes. The rate of fast chemical reactions, including polymerization, is controlled by the rate of the reagent coming to the reaction zone, i.e. the mixing rate (the diffusion rate) of the reacting substances. This fact makes the reaction rate and the reagent conversion depend on the mass of the specimen, the thickness of the reaction zone layer and many other nonchemical factors.

Very fast chemical reactions proceed, as a rule, in the diffusion region and necessarily require the macrokinetic approach to the problem solution. We should note that the name of the science—macrokinetics—for studying the effect of physical processes and hydrodynamics on the chemical reaction run was suggested by Professor D. A. Frank-Kamenetsky [3]. It was most efficiently developed by specialists on burning [4].

Very fast liquid-phase polymerization reactions require the similar macrokinetic approach as well. For them a number of fundamental traditional concepts—for instance, isobutylene polymerization kinetics, and, as a consequence, technological procedure of the production of oligomers and polymers of isobutylene etc.—have to be reconsidered. In some cases, new theoretical and practical approaches must be developed. This fact has created great interest for many years in fundamental studies of liquid-phase electrophilic isobutylene polymerization as a model of very fast reaction. New data continue to create additional interest. For the last few years, it has been clear that the problems of the chemistry and technology of oligo- and polyisobutylene production, as well as of the other superfast chemical reactions and processes limited by mass and heat transfer, are not as simple as they may seem at first sight.

Instead of traditional problems such as process intensification and heat removal, extremely high rates of polymerization reactions give rise to the other and more complicated ones: modeling and controlling of superfast polymerization reactions with the creation of socially and economically beneficial technological procedures for production of polymers with the required characteristics. These characteristics are high specific process efficiency (the maximum output of the product per a unit of the reaction volume) with the minimum capital investment, efficient use of material and power, and control and maintenance of the process, as well as the production of polymers of definite molecular structure and chemical composition.

One should note that the new methodological and experimental approaches revealed on the suitable model of very fast chemical reactions—liquid-phase electrophilic (cationic) isobutylene polymerization—and the original results given later in the book are of general importance and are applicable to other fast liquid-phase polymerization processes—cation, anionic, free-radical ones. They also proved fruitful for many other processes, neither necessarily polymer nor chemical, where mass exchange is of great significance.

In part anticipating the results given later in the book, we should point out that the combination of basically new results in fields of fundamental laws and engineering solutions in the realization of both purely scientific and applied discoveries in very fast liquid-phase processes means that these reactions should be referred to a separate class of chemical reactions, characterized by their own specificity and methodology of investigation. A separate class of oxidation reactions represents a similar example due to the characteristic and original macroscopic features of these reactions.

SYMBOLS

$[A^*]$ the catalyst concentration

D_T the coefficient of the turbulent diffusion

k_i the constant of the initiation rate
k_m the constant of the chain transfer on monomer rate
k_p the constant of the chain propagation
k_t the constant of the termination chain rate
k_{tr} the constant of the transfer chain rate
k_D the constant of the diffusion rate

$\bar{M}_n$ the number-average molecular weight
$\bar{M}_w$ the weight-average molecular weight
$\bar{M}_z$ the z-average molecular weight
M_v the viscosity-average molecular weight
MW the molecular weight
MWD the molecular weight distribution
$[M]$ the concentration of monomer

$\bar{P}_n$ the number average degree of polymerization
$\bar{P}_w$ the weight average degree of polymerization
$\bar{P}_z$ the z-average degree of polymerization

q the heat effect of polymerization

τ_{ch} the time of the chemical reaction
τ_{mix} the average time of the component mixing

V the linear speed of flow

α the coefficient of heat transfer through the wall

λ_T the coefficient of temperature conductivity
μ_T the turbulent viscosity

υ the kinematic viscosity

$\rho_n(j)$ the numerical function of the distribution on the degrees of polymerization

$\rho_w(j)$ the weight function of the distribution on the degrees of polymerization

1 THE MACROKINETICS OF FAST POLYMERIZATION PROCESSES IN LIQUID-PHASE

1.1 THE PECULIARITIES OF VERY FAST POLYMERIZATION PROCESSES

In polymer chemistry there exist processes of very fast polymer synthesis, when the rates of chemical reactions are comparable with or higher than those of the reactant mixing. Such processes, in particular, include some solid-phase reactions in photo- or radiation initiation [5], explosive polymerization of the monomers under shift compression [6], some proceses of cationic polymerization in the presence of sufficiently strong electrophilic catalysts (for instance, the polymerization of styrene, vinyl alkyl ethers, isobutylene etc. in the presence of BF_3, $AlCl_3$, $SnCl_4$ etc.) [1, 2, 7], some reactions of anionic polymerization of vinyl monomers [8], the ionic polymerization of formaldehyde and acetaldehyde [9], the free radical polymerization of ethylene at high pressures [10], complex radical polymerization [11], a number of processes of non-equilibrium polycondensation [12] etc. The constants of chain propagation rates for some of them are given in Table 1.

Among very fast polymerization reactions, one of the most studied in terms of macrokinetics theory, due to general theoretical lucidity and greater practical value, is the electrophilic liquid-phase cationic polymerization of isobutylene [1, 2], which may be taken for the classical model of a fast reaction.

Table 1. The constants of chain propagation rates in fast polymerization based on different mechanisms

Monomer	Solvent	Reaction conditions	k_p, l/mol·s	E_p, kJ/mol	lg A_p l/mol·s	Ref.
		The anionic polymerization				
styrene (polymerization in free polystyrene anions)	tetrahydrofuran	counterion Na^+, 298 K	950	5.9	9.05	[13, 14]
	dimethoxyethane	counterion Na^+, 298 K	4000	~5	8.3	[15]
	tetrahydropyran	counterion Li^+, 296 K	6600	5.1	6.8	[11, 14]
2-vinyl pyridine	tetrahydrofuran	counterion Na^+, 298 K	7300	—	—	[16]
4-vinyl pyridine	tetrahydrofuran	counterion Na^+, 298 K	3500	—	—	[16]
		The cationic polymerization				
isobutyl-vinyl ether	dichlormethane	catalyst $(C_6H_5)_3CSbCl_6$	4000	—	—	[17]
		$C_7H_7SbCl_6$ 273 K	6800	—	—	[17]
isobutylene	—	ZnO, 273 K	$1.5 \cdot 10^8$	—	—	[18]
		γ-radiation, 133–193 K	$3.0 \cdot 10^6$	—	—	[19]
	iso-butane	$AlCl_3$, 243 K	$1.0 \cdot 10^6$	—	—	[20, 21]

The common way to obtain oligo- and polyisobutylene (PIB) with the molecular mass (MW) 112–50,000 in industry is the cationic polymerization of isobutylene (IB) in the presence of $AlCl_3$ in hydrocarbon medium (butane etc.) at the temperature interval of 173–353 K in complex design reactors-mixers with the volume of 1.5–30 m^3 [2, 22].

The application of the PIB depends on the MW, and the polymer product quality in most cases is also dependent on the MWD. This requires strict observation of the technological parameters in obtaining PIB and making the reaction thermostatic.

To provide heat removal and the required output according to thermal and material balances, the reactors have well-developed inner and outer heat-exchanging surfaces (up to 130 m^2 and more) with liquid ethylene or ammonia, highly intensive mixing facilities, capable of linear rates of reaction mass flow of 1–10 m/s and great value of heat exchange in the reactor at an average time interval up to $1.8\text{–}36 \cdot 10^3$ s. The monomer solution (1–10 mol/l) is usually introduced through the upper or bottom part of the reactor, the catalyst solution (10^{-3}–10^{-4} mol/l)-through a relatively thin branch tube, and the product is continuously taken out at the top.

The reaction mass kinematic viscosity $\nu = 10^3$–10^4 kg/m·s and the reactor geometry give the value of the Reynolds number (Re = 10^4–10^5) that predetermines the flow turbulence. In combination with intense mixing (in the classical meaning of this procedure), this should have resulted in the sufficiently uniform distribution of the monomer and the catalyst in the reactor volume with controlled thermal conditions of the reaction, i.e. in isothermal regime of ideal mixing.

Meanwhile regulating the temperature of isobutylene polymerization process presents insuperable difficulties not only in industry, but in the laboratory as well. The diagrams of temperature variation at the local measuring are in practically all cases the regular oscillations of different amplitude (Fig. 1), which is the consequence of the kinetic peculiarities of the cationic polymerization of isobutylene.

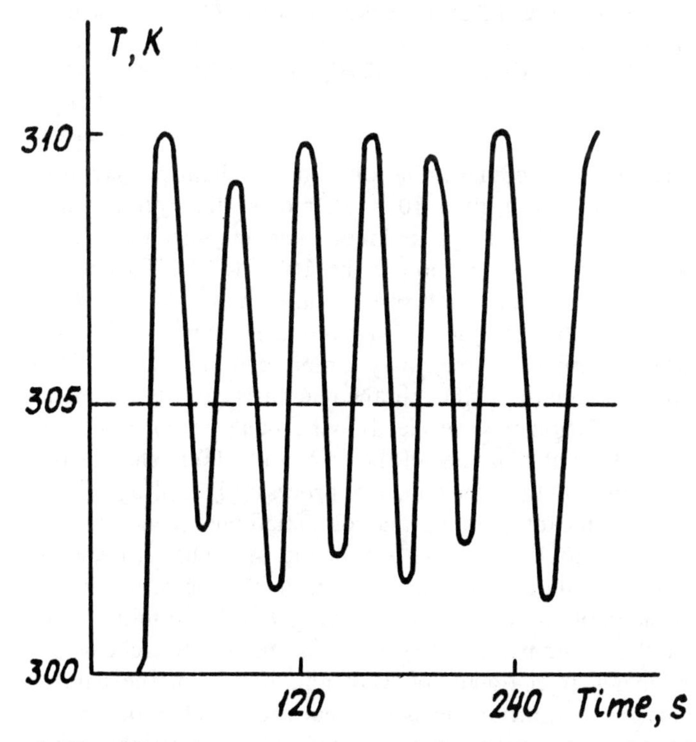

Figure 1. The polymerizate temperature variation in time in real industrial reactors ($V = 1.5\,m^3$, the production of polyisobutylene with $\bar{M}_n = 800$–1000).

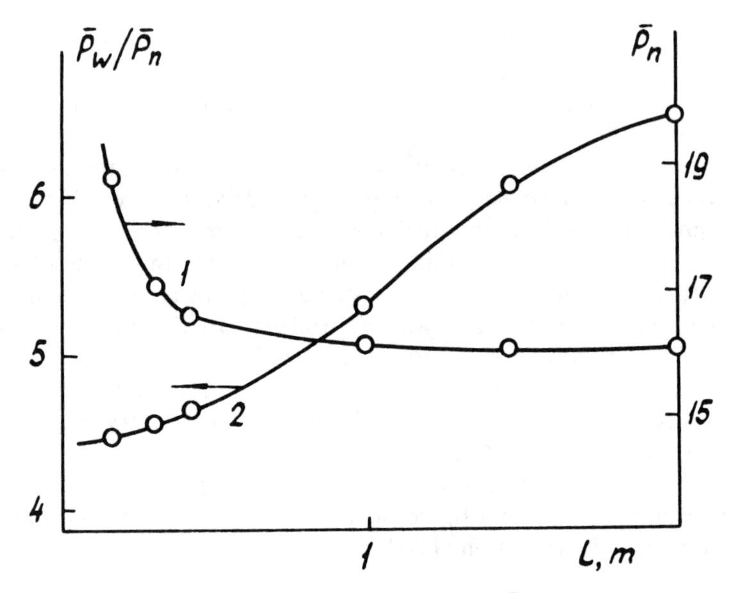

Figure 2. Dependence of the polymerization degree $\bar{P}_n$ (1) and the polydispersion index $\bar{P}_W/\bar{P}_n$ (2) of isobutylene on the reaction zone length ($[M]_0 = 1.9$ mol/l; $[AlCl_3]_0$ in $C_2H_5Cl = 3.7 \cdot 10^{-3}$ mol/l; $T_0 = 248$ K).

The constants of the initiation and the chain propagation rates (k_i, k_p) are not lower than 10^5–10^6 l/mol s., that under real initial concentrations of the catalyst and monomer determine the characteristic time of the chemical reaction to be 10^{-1}–10^{-3} s. This means that isobutylene polymerization proceeds mainly at a distance of less than 1–10 cm from the catalyst entering point. In this case the process is limited by the mixing of the catalyst, monomer and the reaction mass.

The accumulation of heat in the reacting system at the high turbulence of the reaction mass flow leads to the oscillation of temperature due to non-isothermal character of the MW and MWD change along the reaction zone length in the course of process (Fig. 2) [23].

It is significant that the macrokinetic consequence of fast local process of isobutylene polymerization appears to be the dependence of the mean MW on the catalyst and monomer concentration as well as the considerable broadening of the MWD with the initial concentration of the monomer (Fig. 3), whereas the kinetic scheme suggests the independent character of MW and MWD of the catalyst and monomer concentrations in isothermal regime. (The polyisobutylene MW is fully determined by the chain transfer to the monomer.)

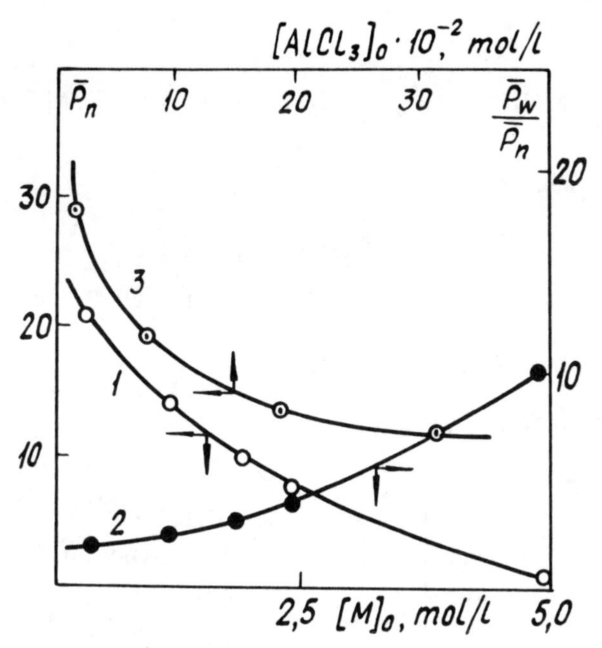

Figure 3. Dependences of $\bar{P}_n$ (1) and $\bar{P}_W/\bar{P}_n$ (2) on the monomer (isobutylene in isobutene) and $\bar{P}_n$ (3) on the catalyst ($AlCl_3$ in C_2H_5Cl) concentrations: 1, 2 – $[AlCl_3]_0 = 3.7 \cdot 10^{-3}$ mol/l; 3 – $[M]_0 = 1.9$ mol/l.

Moreover, the monomer and catalyst solutions, as a rule, are non-uniformly introduced into the reactor. Because of the very high rates of polymerization reaction, they do not manage to become well mixed with the reaction mass. This fact creates additional instability of the reactor work, broadens the MWD and decreases the MW of the final product.

1.2 THE PROBLEMS OF FAST POLYMERIZATION KINETICS

According to the available experimental data, isobutylene polymerization in the presence of metal halogenides (Levis acids) in combination with Bronsted acids proceeds by the following scheme (on the example of using the catalyst on the basis of MeX_n and H_2O): [2, 18].

Formation of the catalyst complex:

$$MeX_n + H_2O \rightarrow MeX_n \cdot H_2O \rightarrow H^{\delta+}, MeX_nOH^{\delta-}$$

Initiation

$$H^{\delta+}, MeX_nOH^{\delta-} + CH_2{=}C(CH_3)_2 \xrightarrow{k_i} (CH_3)_3C^{\delta+}, MeX_nOH^{\delta-} \qquad (I)$$

Chain propagation

$$(CH_3)_3C^{\delta+}, MeX_nOH^{\delta-} + CH_2{=}C(CH_3)_2 \xrightarrow{k_p} (CH_3)_3C{-}CH_2{-}(CH_3)_2C^{\delta+}, MeX_nOH^{\delta-}, \text{etc.} \qquad (II)$$

Chain transfer (mainly to monomer)

$$\sim CH_2C^{\delta+}(CH_3)_2, MeX_nOH^{\delta-} + CH_2 = C(CH_3)_2 \rightarrow$$

$$\xrightarrow{k_m} \sim CH_2{-}C(=CH_2)(CH_3) + (CH_3)_3C^{\delta+}, MeX_nOH^{\delta-} \qquad (III)$$

$$\xrightarrow{k_m} \sim CH{=}C(CH_3)(CH_3) + (CH_3)_3C^{\delta+}, MeX_nOH^{\delta-}$$

Proton transfer to the counterion (a probable reaction) in a number of cases can be disregarded:

$$\sim CH_2C^{\delta+}(CH_3)_2, MeX_nOH^{\delta-}$$

$$\xrightarrow{k_{tr}} \sim CH_2{-}C(=CH_2)(CH_3) + H^{\delta+}, MeX_nOH^{\delta-} \qquad (IV)$$

$$\xrightarrow{k_{tr}} \sim CH{=}C(CH_3)(CH_3) + H^{\delta+}, MeX_nOH^{\delta-}$$

Chain termination with the capture of the counterion fragment

$$\sim CH_2C^{\delta+}(CH_3)_2, MeX_nOH^{\delta-}$$

$$\xrightarrow{k_t} \sim CH_2(CH_3)_2X + MeX_{n-1}OH \qquad (Va)$$

$$\xrightarrow{k_t} \sim CH_2(CH_3)_2OH + MeX_n, \qquad (Vb)$$

the catalyst regeneration with the rapid reinitiation (Reaction V b) being transformed into the reaction of the chain transfer to the counterion.

Thus, for modeling the chemical process of liquid-phase, electrophilic polymerization of isobutylene, the following kinetic scheme should be used:

$$C + M \xrightarrow{k_i} A_1^* \tag{VI}$$

$$A_1^* + M \xrightarrow{k_p} A_2^* \tag{VII}$$

$$A_{n-1}^* + M \xrightarrow{k_p} A_n^* \tag{VIII}$$

$$A_n^* + M \xrightarrow{k_m} A_1^* + \text{Polymer} \tag{IX}$$

$$A_n^* \xrightarrow{k_t} \text{Polymer} \tag{X}$$

The rate of initiation in accord with the experiment is sufficiently high ($k_i \geqslant k_p$); therefore; in the kinetic consideration the concentration of the active sites can be identified with that of the catalyst.

In general, the degree of polymerization according to scheme (I)–(V) depends on the monomer concentration [7]

$$\frac{1}{\bar{P}_n} = k_t + k_m[M]/k_p[M] \tag{1}$$

However, as the chain transfer to the monomer (III) is the basic reaction determining the values of the MW and MWD over the wide range of temperatures, the $\bar{P}_n$ and MWD of the polymer formed in every sufficiently small element of the reaction volume are determined only by temperature and are independent of the catalyst and monomer concentrations.

$$\bar{P}_n = \frac{k_p}{k_m} = \frac{k_p}{k_m} \exp(E_m - E_p)/RT \tag{2}$$

$$\rho_n(j) = \frac{1}{\bar{P}_n} \exp(-j/\bar{P}_n), \tag{3}$$

where $\rho_n(j)$ is the numerical function of the distribution on the degrees of polymerization; j is the degree of polymerization.

Isobutylene is characterized by its extremely high reactivity with respect to the cationic agents. This causes a very high rate of polymerization accompanied by evolution of a considerable quantity of heat, which, as a rule, cannot be removed from the reaction zone. It is the high order of the numerical values of the chain propagation rate constant (k_p) in electrophilic polymerization of isobutylene that makes it possible in principle to refer the reaction of cationic polymerization of IB to extremely fast polymerization processes. The values of k_p are below the diffusion limit. In particular, k_D for the most monomers with viscosity of 0.5 kPa·s at 300 K is about 10^{10} l/(mol·s). However, in the formation of a viscous product, especially in the low-temperature polymerization of isobutylene, the values of k_D are markedly decreased; for instance, k_D for the system of isobutane-polyisobutylene is about 10^6 l/mol·s [24], and the kinetic constants k_p in many cases may be even higher k_D.

With the extremely high rates of polymerization and the exothermicity of the process ($q = 54.0$ kJ/mol), even a very slow influx of the initiator and instant mixing are not enough to remove heat evolved in the reaction. Only the use of the maximum diluted monomer solutions (about 0.01–0.02 mol/l), which, naturally, present no interest for the real processes of polymerization of isobutylene, can achieve conditions close to isothermal [1, P. 144]. In a general case, the reaction of isobutylene polymerization starts before the initiating particles have diffused sufficiently far. Even high-speed filming (about 3,000 frames per second) did not make it possible to establish the period between the $AlCl_3$ solution drop hitting the surface of isobutylene (at 195 K) and thc fact of polymerization [1]. It follows from the above that in these and many other ionic and non-ionic systems, the uniformity of disributing reagents and temperature in the volume is not secured. This means that in practice the MW of the polymers formed appears to be markedly lower and the MWD appears to be wider; that is, the real processes of the cationic polymerization of isobutylene and other similar extremely fast chemical reactions are difficult to control. These circumstances require searching and elaborating new approaches to the kinetic study of fast polymerization processes, new methods of evaluating the kinetic and other characteristics of the corresponding reactions, and effective systems of controlling the processes indispensably using the equations of the chemical kinetics, heat transfer, diffusion, and convection.

1.3 EXPERIMENTAL MODELING OF FAST POLYMERIZATION PROCESSES

The special methodological requirements for studying fast polymerization processes have been revealed in the analysis of variations in the real values of the MW and the MWD of the polymer products formed

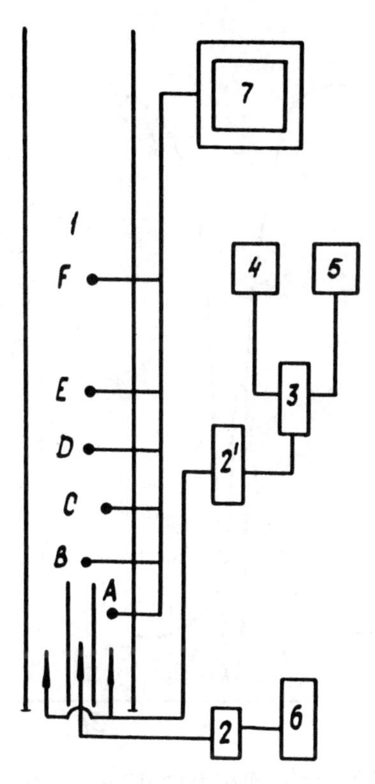

Figure 4. Scheme of the polymerization device: 1 – reactor; 2,2′ – pumps; 3 – vessel for monomer and solvent mixing; 4,5,6 – vessels for monomer, solvent and catalyst; 7 – recorder; A–F are the points of sampling.

depending on the conditions of the reaction. In [25] researchers used a laboratory model device for the production of polyisobutylene in a flow (Fig. 4), that allowed control of the linear rates of introducing reagents, measurement of the temperature in the preassigned points of the reaction volume, and control of its change.

The solutions of isobutylene (1,7 mol/l) and the catalyst ($C_2H_5AlCl_2$, $1.4 \cdot 10^{-2}$ mol/l) in heptene were entered with the constant speed of the flow ($1 \cdot 10^{-2}$ m/s and higher). The MW and MWD of the polymer formed were measured by a gel chromatograph supplied with two detectors: a refractometer and a specific double-bond analyzer (DBA) [26, 27].

For the obtained products (PIB) with chain end double bond, DBA-chromatograms differ from the gel-chromatograms because the

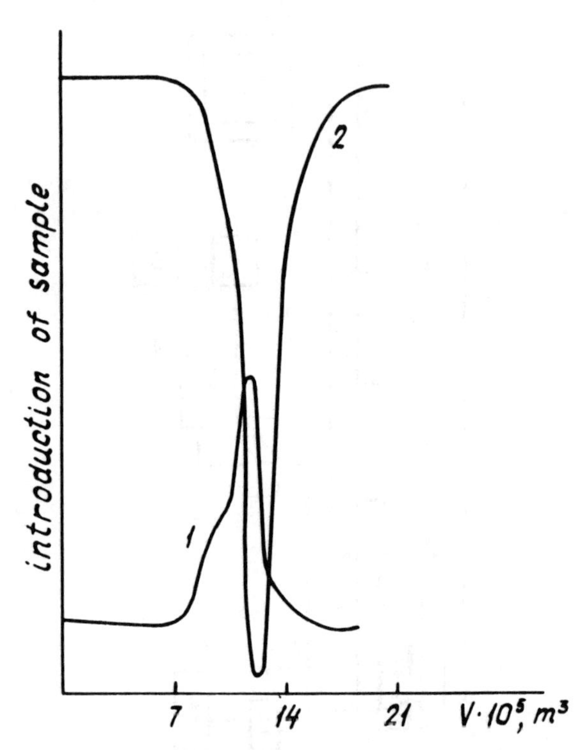

Figure 5. Gel-(1) and ADB-chromatogramme (2) of polyisobutylene ($\bar{M}_n = 860$).

proportion of unsaturation in the low-molecular field is higher than that in the high-molecular one. (The typical curves are given in Fig. 5.) As practically every molecule of polyisobutylene contains one chain end double bond, the use of the specific double-bond analyzer allows for the PIB oligomers to obtain calibration by simple joint processing of the two chromatograms (Fig. 6). The conditions of the specimen analysis were: the eluent-n-heptane, the elution rate is $2.3 \cdot 10^{-8}\,m^3/s$, columns ($d = 8 \cdot 10^{-3}\,m$, $1 = 1.2\,m$) with styrogel sorbent $2 \cdot 10^2$, $5 \cdot 10^2$, $1 \cdot 10^3$ A, temperature is 293 K, the rate of the ozone-oxygen mixture supply is $0.74 \cdot 10^3$ l/s, the temperature in the ozonization cell is 273 K.

The mean values of the MW were determined with DBA and the joint processing of the DBA- and gel-chromatograms by the methods of iodometry and CHEM*. The calculations according to the DBA method were made by the formula:

$$MW = 1/D,$$

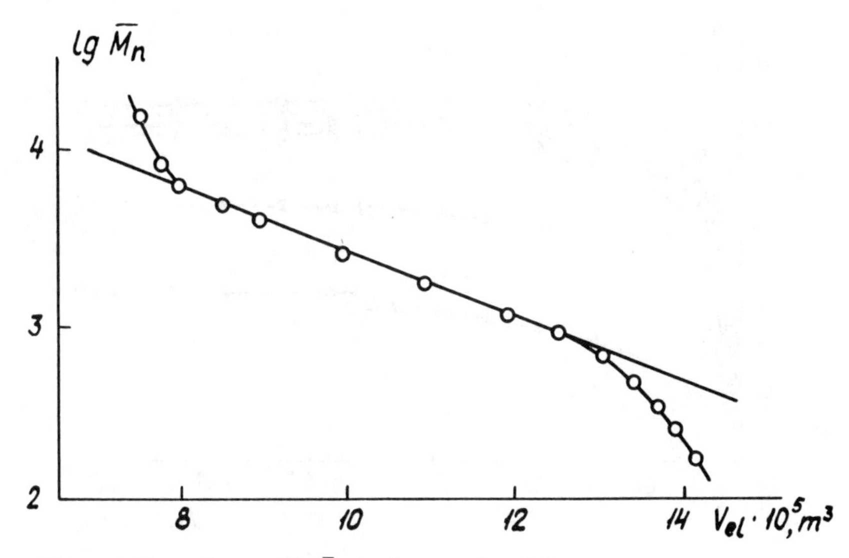

Figure 6. Dependence of lg $\bar{M}_n$ on the quantity of eluents.

where

$$D = \frac{C_{st} \cdot V^{st} \cdot S_{sp} \cdot V_{sol}}{S_{st} \cdot V^{sp} \cdot q_{sp}} \tag{4}$$

Here D is the degree of unsaturation ($C = C$ per·g of polyisobutylene), C_{st} is the stilbene concentration (mol/sm^3), V^{st}, V^{sp} and V_{sol} are the volumes of the stilbene, specimen and solution probes, S_{sp} and S_{st} are the areas of the specimen and stilbene peaks, q_{sp} is the probe charge (g).

The temperature variation in different points of the reaction zone testifies to the process proceeding at the highest rate at the catalyst entering point (point B, Fig. 7). The lowest rate of the reaction is observed near the reactor walls (point C).

The data, summarized in Table 2, indicate that the reaction in the experiment conditions mainly runs (by 50–70%) in the reagent entering zone during 1–2 s. (point B).

The temperature of the reaction mass appears to be variable and to depend on the concentration of the reagents, isobutylene in particular (Table 2, experiments 9–12). The decrease of the polymer MW and the reduction of the monomer conversion have been registered simultaneously with the increase of the monomer concentration in the initial mixture. The experimental data observed are caused by the increase in

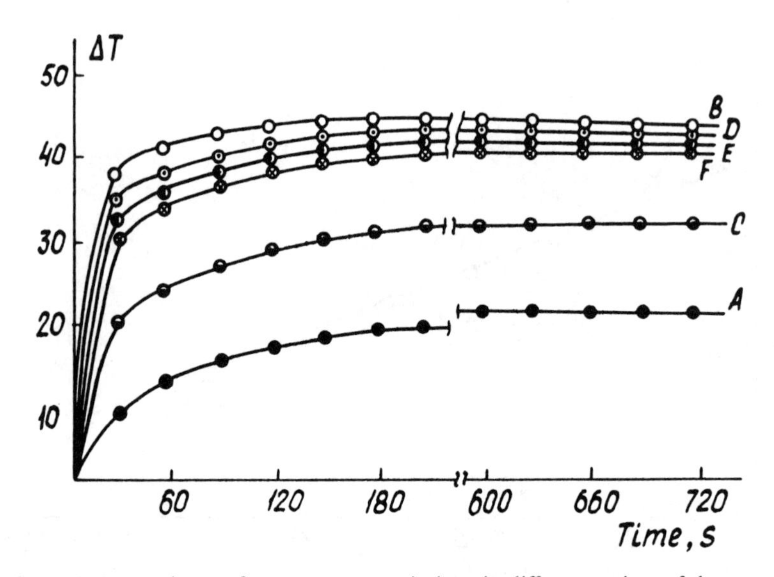

Figure 7. Dependence of temperature variations in different points of the reaction zone on time; A–F are the points of sampling (See Fig. 4).

Table 2. The dependence of the MW of the product and monomer conversion on the reaction zone length and the monomer concentration

Number of experiment	Points of sampling	ΔT**	Reaction zone length, l, m	Conversion, % wt.	$\bar{M}_n$ of polymer	
					CHEM	DBA
1	A	35	0.02	56.2	960	950
2	B	36	0.10	72.0	940	935
3	C	37	0.30	86.3	700	690
4	D	40	0.48	98.4	510	500
5	A	32	0.02	48.9	1250	1210
6	B	35	0.10	70.4	1000	980
7	C	37	0.30	82.6	840	840
8	D	40	0.48	98.6	560	550
9	C	24	0.30	81.0	1500	1480
10	C	30	0.30	71.2	1210	1150
11	C	37	0.30	70.3	950	930
12	C	44	0.30	69.2	820	820

Monomer concentration 12%wt, rate of reagent supply 0.01 m/s
** The initial temperature 273 K

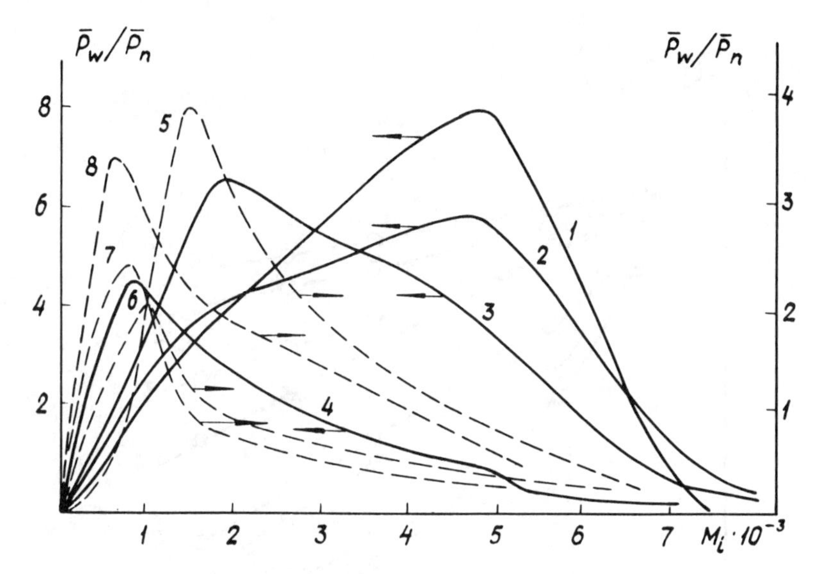

Figure 8. Dependences of polyisobutylene MWD on the reaction zone length (1–4) and monomer concentration (5–8): 1–4 – correspond to the results of experiments 5–8, 5–8 – correspond to the results of experiments 9–12 (Table 2).

quantity of the heat evolved in the polymerization, natural loss in the heat-exchange and by the growing *CHEM*-Condensation heat effect measurement role of the reaction of terminating the material chain by chain transfer to monomer. When the reaction mass is moved away from the catalyst entering point (increase in the conversion up to 100%) the isobutylene polymerization rate is changed due to the other, than in point B, gradients of the monomer temperature and concentrations, that determine the dependence of the MW (Table 2) and MWD values (Fig. 8) on the reactor length. The curves of the differential MWD differ with variation in the isobutylene concentration, too.

The MWD functions in coordinates lg $\rho_n(j)$ from j for both cases are given in Fig. 9. One can clearly see the MWD broadening due to the appearance in the product of considerable amounts of the low-molecular fraction, along with moving away from the catalyst entering point as well as with the increase in the monomer concentration of the reaction mixture. Hence, the gradients of temperature and various rate fields in the polymerization zone determine the increase in nonuniformity of the polymer product by the MW. In other words, in the topochemical aspect the isobutylene polymerization is referred to essentially fast reactions and may be termed a “torch” in the reaction of burning characterized by

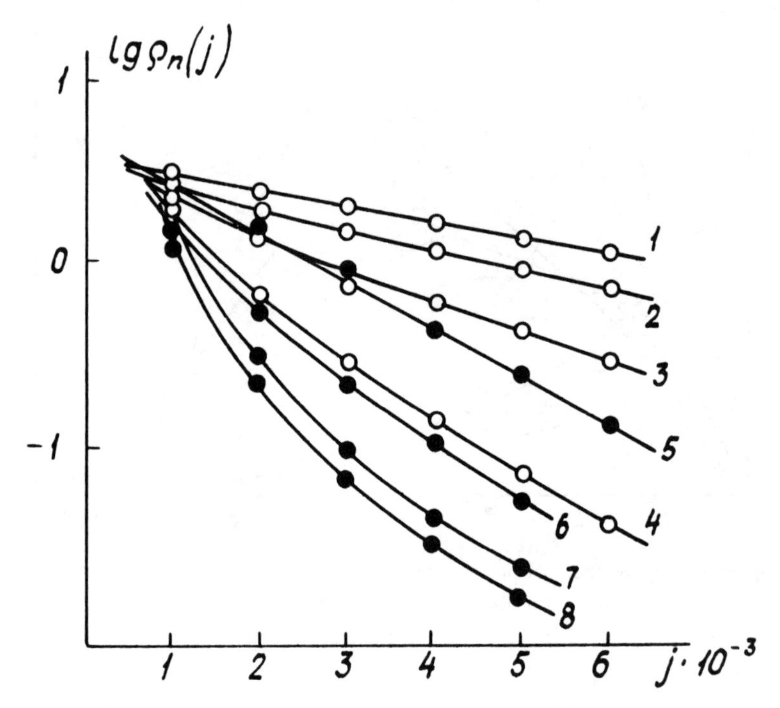

Figure 9. Semilogarithmic anamorphosis of the function MWD polyisobutylene from the reaction zone length (1–4) and monomer concentration (5–8); l, m: 1–0.02; 2–0.1; 3–0.3; 4–0.48 at $[M]_0 = 1.9$ mol/l; $[M]_0$, mol/l: 5–0.8; 6–1.6; 7–1.9; 8–2.4 at $l = 0.3$ m.

different temperature zones, reagent concentrations, process rates, etc., along the reaction zone coordinates.

1.4 THE COMPUTATION AND MODELING OF FAST POLYMERIZATION REACTION

The mathematical modeling of fast polymerization in a flow (stream) is based on the kinetic scheme of the isobutylene polymerization reactions (VI)–(X) and the correlations (2) and (3) [28, 29]. It has been taken into account that in polymerization in ideal mixing reactors, high linear rates of reaction mass movement (up to 10 m/s) are often secured. It is expedient to use the method of polymerization in a flow (stream) from the several points of view, noted in [30]; it is the most convenient way to study experimentally extremely fast chemical reactions and the best way for running the reaction with the maximum intensity in minimum volume and time. The investigation of the kinetics of the reaction in a stationary flow with intense mixing is basically simpler than investigation of chemical reactions proceeding in a closed volume; the former ones, including extremely fast polymerization reactions, suggest the easier and more convenient verification of the theoretical modeling by polymer product sampling at different points along the reaction volume length and the definition of the characteristic parameters of the polymer that was formed.

The reaction zone of the apparatus for computation and modeling is selected from the laboratory model (Fig. 4) with the method of scale transfer; i.e., the same correlations of the apparatus geometry, the rates of the reagent supply and also the principle of the catalyst influx were followed.

The high rates of the flow in the reaction zone (1 to 10 m/s) provide the turbulent mixing of the catalyst solution ($[A^*]_0 = 10^{-4}$–10^{-1} mol/l), the polymer formed and the solvent used (isobutane, methylene chloride etc.). The values of the Reynolds number (Re), calculated for the given linear rate of the flow, its density $(0.5–1.0)\cdot 10^3$ kg/m^3, the coefficient of dynamic viscosity ($5\cdot 10^{-4}$ kg/m·s), and the diameter of the reactor (0.1 m), appear to be 10^4 and higher. The high turbulence is the result of the reaction (in polymerization the contraction is about 30–40% wt). The reaction mass may boil because of the generation of considerable amounts of heat. Therefore, the coefficient of turbulent diffusion equal to heat conductivity ratio can be taken in the first approximation as coefficients of mass- and heat transfer

$$D_T = \frac{\lambda_T}{\rho \cdot c} \tag{5}$$

The theoretical calculations of the hydrodynamics of such systems are usually extremely difficult and require introducing many empirical parameters. Meanwhile, for some model systems, close to the real ones, calculations are quite possible. In the strict sense, the similar systems should be described by the effective coefficient of turbulent diffusion D_T, changeable in space. However, as the special evaluations have shown, D_T variations in the total reaction volume in open systems are not great (less than in twice) and are still smaller in the effective chemical reaction zone. In fact, it should be borne in mind, that the efficiency of mixing depends heavily on the reaction volume configuration, on the system of the catalyst supply relative to the direction of flows, on the energy of flow turbulization, etc. The calculations have shown that the D_T value (excluding the contraction due to polymerization and boiling of the reaction mixture components), depending on the reaction zone profile at the rational values of the energies of turbulizations, is in the limits of 10^{-2}–10^{-3} m^2/s (for the laminar flow-10^{-9} m^2/s respectively); when we take into account boiling and contraction, the value of D_T is still higher. Thus, in calculating and modeling the fast reaction of isobutylene polymerization D_T by the reaction volume and correspondingly the coefficient of temperature, conductivity λ_T has been supposed to be constant, the D_T values being varied in the interval of 10^{-3} to 10^{-1} m^2/s.

Since the initiation rate in the electrophilic polymerization of isobutilene is sufficiently great ($k_i \geqslant k_p$) and the initial equations do not change if we suggest the initiation be relatively slow and the active site concentration be stationary and the main reaction determining the MW and MWD of the polymer formed within a wide range of temperatures and especially at high temperatures be the chain transfer to monomer, then the degree of polymerization $\bar{P}_n$ and MWD $\rho_n(j)$ in every small element of the reaction volume is determined only by temperature Eq. [(2) and (3)]. Therefore, the NW and MWD variations of the product formed in the isobutylene polymerization as a model of an extremely fast reaction reflect the temperature field in the reaction zone.

According to the above and to the experimental data, the kinetic equations describing the variations in monomer concentration andthose of active sites and the temperature in fast polymerization of isobutylene in the flow with the axial symmetry have the from:

$$\frac{\partial[M](x,r)}{\partial t} = D_T\frac{\partial^2[M]}{\partial r^2} + \frac{D_T}{r}\frac{\partial[M]}{\partial r} + D_T\frac{\partial^2[M]}{\partial x^2} - V\frac{\partial[M]}{\partial x} - k_p^0[M][A^*]\cdot\exp(-E_p/RT) \quad (6)$$

$$\frac{\partial[A^*](x,r)}{\partial t} = D_T\frac{\partial^2[A^*]}{\partial r^2} + \frac{D_T}{r}\frac{\partial[A^*]}{\partial r} + D_T\frac{\partial^2[A^*]}{\partial x^2} - V\frac{\partial[A^*]}{\partial x} - k_t^0[A^*]\cdot\exp(-E_t/RT) \quad (7)$$

$$\rho c\frac{\partial T(x,r)}{\partial t} = \lambda_T\frac{\partial^2 T}{\partial r^2} + \frac{\lambda_T}{r}\frac{\partial T}{\partial r} + \lambda_T\frac{\partial^2 T}{\partial x^2} - V\cdot\rho\cdot c\frac{\partial T}{\partial x} + q\cdot k_p^0[M][A^*]\cdot\exp(-E_p/RT) \quad (8)$$

In the case of the boundary conditions at the central catalyst supply in the flow of the monomer solution:

$$T(-d,r)=T_0 \qquad \text{at } O<r<R$$

$$[M](-d,r)=\begin{cases}[M]_0 & \text{at } R>r>r_0\\ O & \text{at } r<r_0\end{cases} \tag{9}$$

$$[A^*](-d,r)=\begin{cases}O & \text{at } R>r>r_0\\ [A^*]_0 & \text{at } r<r_0\end{cases} \tag{10}$$

$$\frac{\partial[M](x,r)}{\partial r}=\frac{\partial[A^*](x,r)}{\partial r}=\frac{\partial[A^*](x,O)}{\partial r}=\frac{\partial[M](x,O)}{\partial r}$$

$$=\frac{\partial T(x,O)}{\partial r}=O \tag{11}$$

$$\frac{\partial T(x,R)}{\partial r}=\alpha\{T_0(x,R)-T_1\} \tag{12}$$

Here x and r are the reaction zone coordinates by length and radius; T-temperature, T_0 and T_1 are the temperature of the reagents introduced and that of outer thermostating; α – the coefficient of heat transfer through the reactor wall; q – the heat effect of polymerization (for isobutylene -54.0 kJ/mol); $k_p=k_p^0 \exp(-E_p/RT)$, $k_t=k_t^0 \exp(-E_t/RT)$, $k_m=k_m^0 \exp(-E_m/RT)$ are the rate constants of the chain propagation, termination and transfer to monomer respectively; k_p^0, k_t^0, k_m^0 and E_p, E_t, E_m are the coefficients and energies of activation of these reactions.

Taking into account the stationary character of the process, i.e.,

$$\frac{\partial M(x,r,t)}{\partial t}=\frac{\partial A^*(x,r,t)}{\partial t}=\frac{\partial T(x,r,t)}{\partial t}=O$$

and neglecting the diffusion transfer along the axis X as compared to the linear one with the rate V, the solution of quasilinear parabolic equations (6)–(8) can be simplified by dropping the terms:

$$D_T\frac{\partial^2[M]}{\partial x^2},\ D_T\frac{\partial^2[A^*]}{\partial x^2},\ \lambda_T\frac{\partial^2 T}{\partial x^2}$$

As a result, the computation of the mathematical model brings about the solution of the boundary-value problem for the system of equations (6)–(8) by means of the implicit difference scheme.

In addition to the computation of the fields of concentration $M(x, r)$ and $A^*(x, r)$ as well as to that of the reaction rate fields $(k_p[M][A^*])$, the MWD and the average MW of the polymers formed have been calculated. The mass function of the distribution $\rho_W(j)$ by the degree of polymerization at the outlet of the reactor with the length l and the moments of the distribution function I_0–I_3 have been calculated by the formulas:

$$\rho_W(j) = \int_{-d}^{l} \int_0^R \frac{j}{\bar{P}_n^2} \exp(-j/\bar{P}_n) k_p^0 [A^*][M] \exp(-E_p/RT) \cdot 2\pi r dx dr, \tag{13}$$

where

$$\bar{P}_n = k_p/k_m = k_p^0/k_m^0 \exp[-(E_p - E_m)/RT]$$

$$I_0 = \int_0^\infty \frac{\rho_W(j)}{j} dj = \int_0^R \int_{-d}^{l} k_m^0 [A^*][M] \exp(-E_m/RT) \cdot 2\pi r dr dx$$

$$= \int_0^R \int_{-d}^{l} \frac{1}{\bar{P}_n} k_p^0 [A^*][M] \exp(-E_p/RT) \cdot 2\pi r dr dx \tag{14}$$

$$I_1 = \int_0^\infty \rho_W(j) dj = \int_0^R \int_{-d}^{l} k_p^0 [A^*][M] \exp(-E_p/RT) \cdot 2\pi r dr dx \tag{15}$$

$$I_2 = \int_0^\infty j \rho_W(j) dj = \int_0^R \int_{-d}^{l} 2\bar{P}_n k_p^0 [A^*][M] \exp(-E_p/RT) 2\pi r dr dx$$

$$= \int_0^R \int_{-d}^{l} [(k_p^0)^2/k_m^0][A^*][M] \exp[-2E_p - E_m)/RT] \cdot 4\pi r dr dx \tag{16}$$

$$I_3 = \int_0^\infty j^2 \rho_W(j) dj = \int_0^R \int_{-d}^{l} 6\bar{P}_n^2 k_p^0 [A^*][M] \exp(-E_p/RT) \cdot 2\pi r dr dx$$

$$= \int_0^R \int_{-d}^{l} 6 [(k_p^0)^3/(k_m^0)^2][A^*][M] \exp[-(3E_p - 2E_m)/RT] \cdot 2\pi r dr dx \tag{17}$$

$\bar{P}_n = I_1/I_0$, $\bar{P}_W = I_2/I_1$ and $\bar{P}_z = I_3/I_2$ have been calculated at the outlet of the reactor.

The integrals $\int_0^R \int_{-d}^{l} f(x,r) dr dx$ have been calculated by the scale available. The numerical values of E_p and E_t for the fast processes of cationic polymerization of isobutylene are not great [19]. Therefore, the calculations of temperature dependences k_p, k_t and D_T have been ignored.

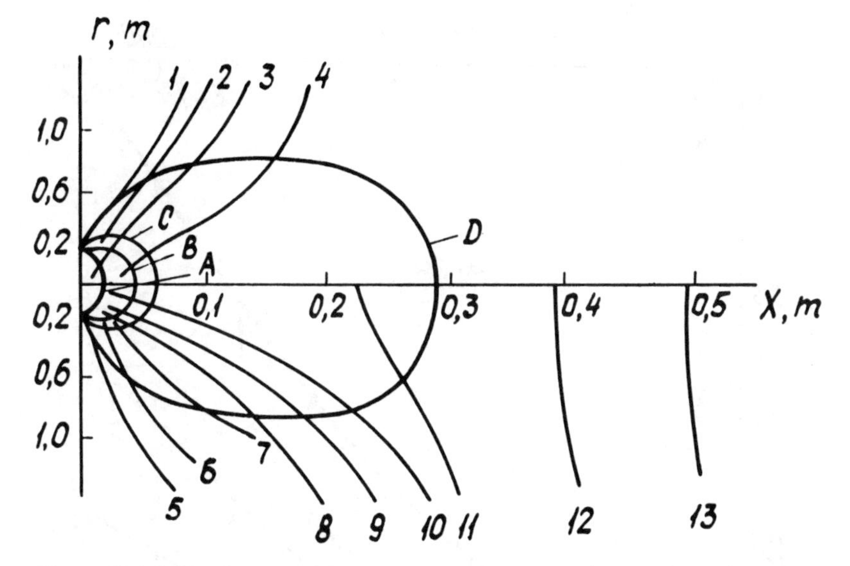

Figure 10 Fields of temperatures, monomer concentrations and catalyst concentrations: $k_p = 10^5$ l/mol·s; $k_t = 1\,s^{-1}$; $[M]_0 = 1$ mol/l; $A_0^* = 10^{-2}$ mol/l; $D_T = 1\,m^2/s$; $\alpha = 0$; T, K: 310(1); 313(2); 320(3); 330(4); $[M]_0$ mol/l: 0.9 (5); 0.7 (6); 0.5 (7); 0.3 (8); 0.2 (9); 0.15 (10); 0.085 (11), 0.035 (12), 0.016 (13); $[A^*]_0$, mol/l: $2\cdot10^{-3}$ (A); $1\cdot10^{-3}$ (B); $5\cdot10^{-4}$ (C); $1\cdot10^{-4}$ (D).

The fields[1] of temperature, monomer and catalyst concentrations are given in Fig. 10. It is clear that the process of isobutylene cationic polymerization proceeds mainly as well as during the experiment at the inlet of the catalyst in the reaction zone at mixing the catalyst with the monomer solution. As is characteristic of fast chemical processes, the temperature and rate in the isobutylene polymerization zone prove to be variable and dependent on the initial concentrations of the reagants, on the D_T value and the coefficient of heat transfer through the wall α. Though the polymerization rate maximum is observed near the catalyst entering zone, the reaction proceeds rather far along the axis x, resulting in changes of the yield and properties of the polymer as it moves away from the catalyst entering point.

The polymer formation in different points of the reaction volume (at different correspondingly temperatures) results in the MWD broadening as compared with the most probable $[\rho_n(j) = 1/\bar{P}_n \exp(j/P_n)]$ which is typical for isothermal conditions.

---

[1] Field is assumed to be variation of the studied parameter along the reaction zone coordinates expressed in relative units.

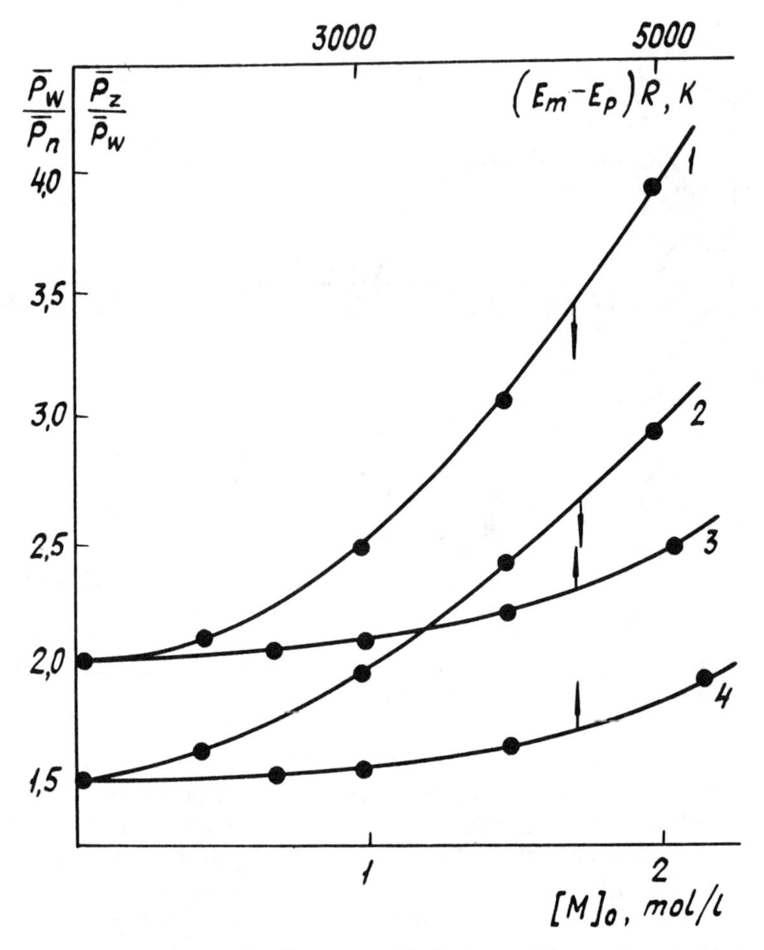

Figure 11. Dependences $\bar{P}_w/\bar{P}_n$ (1, 3) and $\bar{P}_z/\bar{P}_w$ (2, 4) on monomer concentration and $(E_m - E_p)/R$ ($k_p = 10^5$ l/mol·s; $k_d = 1\ s^{-1}$; $[A^*]_0 = 10^{-2}$ mol/l; $D_T = 1\ m^2/s$; $\alpha = 0$; $l = 1$ m, for 1, 2- $(E_m - E_p)/R = 5000$ K); for 3, 4-$[M]_0 = 1$ mol/l.

Since the mean MW and MWD are determined by the reaction of the chain transfer to monomer, the main factor affecting these characteristics is the difference between the activation energies of the reactions of chain transfer and chain propagation E_m–E_p. The dependences $\bar{P}_w/\bar{P}_n$ and $\bar{P}_z/\bar{P}_w$ on E_m–E_p are given in Fig. 11. The marked deviation of the MWD value from the most probable one is caused by the difference of at least several units between the values of E_m and E_p, which on the whole is in agreement with the experimental data for isobutylene polymerization.

The increase of the α coefficient, with the other process parameters being constant, results in a certain MWD narrowing, which corresponds to the temperature field. But only in the case when the reaction vessel is small and D_T and λ_T values are great can heat transfer exert notable influence on the reaction through the impenetrable side wall. The temperature field in the reaction zone is determined by the rate of the process, the amount of heat evolved, and consequently, the monomer and catalyst concentrations.

The monomer concentration effect on the MWD parameters is also given in Fig. 11. The MWD is markedly broadened with the increase of the considerable amount of the low-molecular fraction. The analogous dependences are observed in variation of the catalyst concentration. Besides the expected increase of the polymer yield, the increase of the catalyst concentration results in increases of temperature and of the temperature gradient in the reaction zone and, consequently, in reducing the meen MW and broadening MWD (the increase of $\bar{P}_w/\bar{P}_n$ ratio).

The analysis of fast polymerization reactions has shown that the effects revealed in mathematical modeling are identical to those being observed during the experiments on the example of the cationic polymerization of isobutylene. The deterioration of the polymer quality is the important consequence of the nonisothermicity of the process, the outer thermostating generally being not sufficiently effective and thus limiting the use of dilatometry and many other classical and special experimental methods of studying the process kinetics.

Hence, obtaining the polymer products with the MWD close to the typical one for isothermal processes in a single apparatus requires in general the reduction of the reaction zone to the fields of comparatively low monomer conversion. This in its turn limits the size of the reaction zone itself (the apparatus size).

Computations in conditions approximating the real ones demonstrate that in the catalyst and monomer contact the fields of temperature variation as well as those of the monomer and catalyst concentration are being formed (Fig. 12). In this case, in standard industrial reactors of ideal mixing, the conversion in one run is 25 – 10%wt (which fairly corresponds to the experimental data), and the front of the reaction propagation is characterized by the formation of the local zone ("torch"). The front of propagation of isobutylene polymerization is less than the reaction zone volume which can result in the formation of the monomer slip zone.

The presence of the "torch" conditions—the gradients of temperature, the monomer and catalyst concentrations, the monomer slipping along the outer walls of the reactor—leads to reduction of the poly-

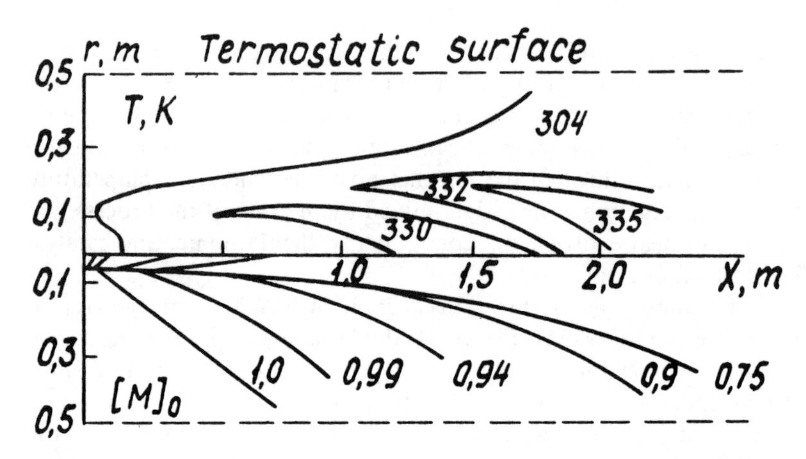

Figure 12. Fields of temperature (T) and monomer concentration ($[M]_0$), in fast polymerization: $[M]_0 = 1$ mol/l; $[A^*]_0 = 4.5 \cdot 10^{-3}$ mol/l; $D_T = 0.025$ m^2/s; $R/r = 10$; $k_t = 20\,s^{-1}$.

isobutylene yield at a run, excess reaction time and, as a consequence, declines in efficiency and productivity of type of volumetric reactors available in industry. Several conditions mark a new macrokinetic peculiarity of isobutylene polymerization. These include the rapid and local character of the reaction, the presence of "torch" conditions, when the reaction zone is not broad enough to reach the inner and outer heat exchange surfaces of the reactor. In this case, total heat removal appears ineffective and the cooling devices can only cool down the reagents as they approach the site of introducing the catalyst and the monomer, which is not effective at all.

2 THE LAWS OF FAST POLYMERIZATION PROCESSES IN LIQUID-PHASE IN FLOWS

2.1 A FEW MACROKINETIC REGIMES

The analysis of experimental and computational results of fast polymerization processes in the example of isobutylene polymerization established a marked influence of the reaction zone geometry—its radius *R* and the length l—on the kinetic parameters of the process and the depth of the monomer conversion (Figs. 13, 14) [31–33].

Apparently, one of the most essential results is the effect of the reaction volume geometry on the molecular characteristics of the products formed in extremely fast liquid-phase polymerization processes. In particular, the examples of Fig. 14 and Table 3 show the influence of the geometric parameters *R* and 1 of the reaction volume on the polymer MW and MWD.

In the topochemical aspect three macrokinetic types of the process—A, B, and C—can be differentiated (Fig. 15).

When the radius *R* is small (type A), the reagent mixing is sufficient, the active centres A* are comparatively regularly distributed (Fig. 15b) and, as a consequence, the reaction temperature is regularly distributed along the reaction zone radius (Fig. 15a). The surfaces of the monomer equal concentrations constitute the planes perpendicular to the reaction zone axis. All these conditions lead to the high (up to 100% wt) conversion of the monomer and the regime of quasi-ideal displacement in high-turbulent flows.

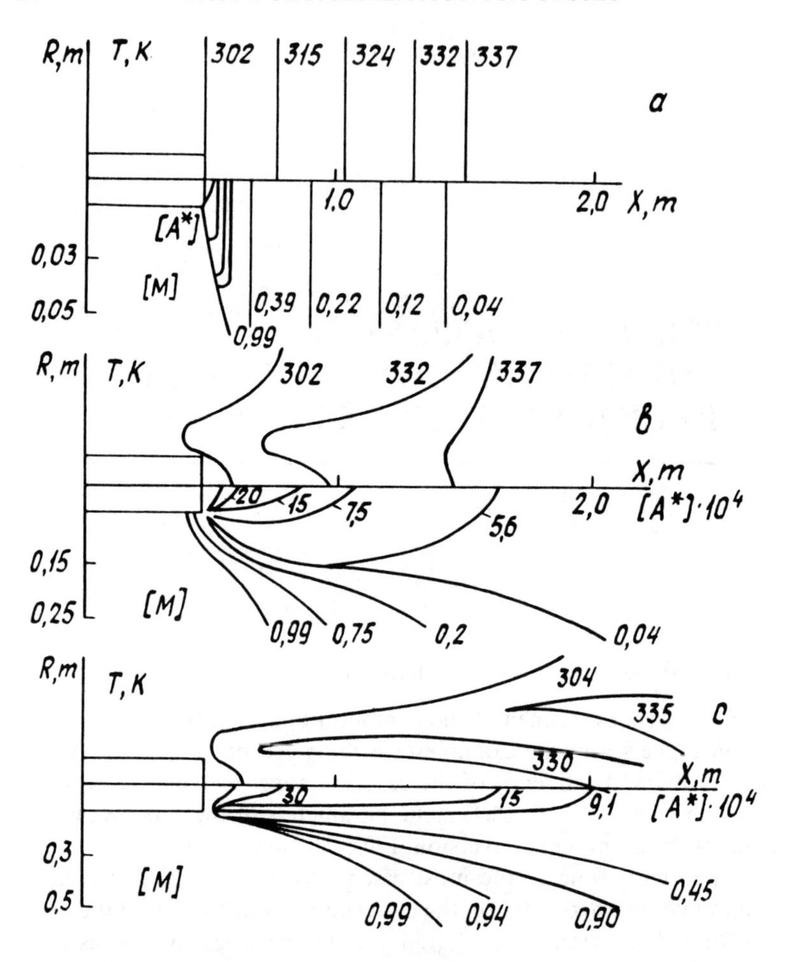

Figure 13. Dependence of fields of temperatures, monomer concentration and catalyst concentration on the radius of the reaction zone R, m: $[M]_0 = 1$ mol/l; $[A^*]_0 = 4.5 \cdot 10^{-3}$ mol/l; $D_T = 0.025\,\mathrm{m^2/s}$; $R/r = 5$; $k_p = 10^5$ l/mol·s, $k_t = 20\,\mathrm{s^{-1}}$. R, m: a -0.05; b -0.25; c -0.50.

Another limiting case (the local "torch" regime) is realized at the relatively great values of R (type C). The active sites A* are destructured having no time to diffuse to the circumferential zones of the reaction volume, which thus constitute the slip zones of the monomer that has not reacted yet (Fig. 15e). This process is accompanied by the formation of the specific fields of the monomers, active sites and temperatures (Fig. 15f), that cause the occurence of the "torch." The size of the torch is determined by the correlation of two competing processes, namely the

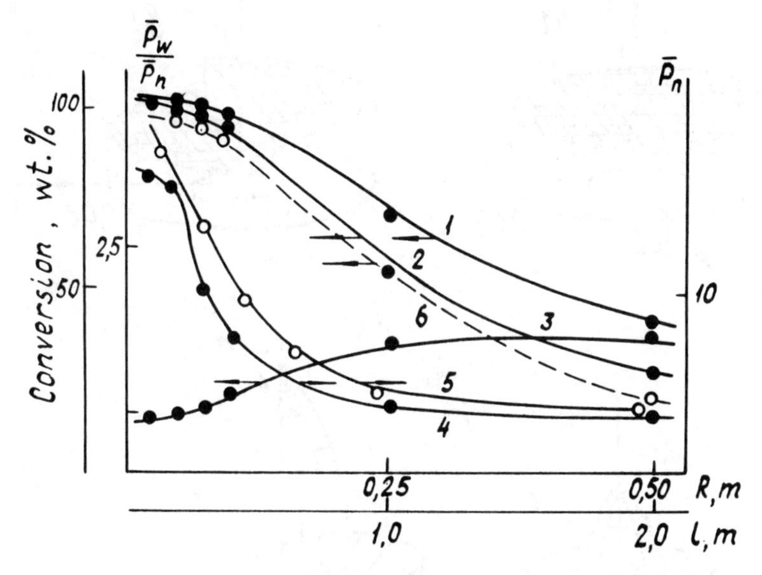

Figure 14. Dependence of the monomer conversion, %wt (1, 2, 6), the polydispersion index $\bar{P}_w/\bar{P}_n$ (3) and the degree of polymerization $\bar{P}_n$ (4, 5) on the reaction zone radius (R, m) and length (l, m) due to calculation (1–5) and experimental dates (6) at: $V = 5$ m/s; $D_T = 0.025\,m^2/s$; 248 K; $k_p = 10^5$ l/mol; k_t, s^{-1}: 15(1); 20(2–5); $[M]_0$, mol/l: 1 (1–5); 1.9 (6); $[A^*]_0 \cdot 10^3$ mol/l ($AlCl_3$ in C_2H_5Cl): 4, 5 (1–5); 3.7 (6).

Table 3. The dependence of the monomer conversion depth, the degree of polymerization $\bar{P}_n$ and the polydispersity index $\bar{P}_w/\bar{P}_n$ on the reaction zone radius R ($k_p = 10^5$ l/mol·s; $k_t = 20\,s^{-1}$; $[M]_0 =$ 1 mol/l;$[A^*] =$ 0.0045 mol/l; $D_T = 0.025\,m^2/s$)

The type of reaction	R, m	Conversion % wt	$\bar{P}_n$**	$\bar{P}_w/\bar{P}_n$**
A	0.01	100	13	2.0
	0.03	100	13	2.0
	0.05	100 (97.7)*	12 (30)	2.1 (3.1)
	0.08	99.3	10	2.1
B	0.10	90.0 (90.0)*	8 (21)	2.2 (3.7)
	0.25	65.0	6	2.4
C	0.50	32.0 (29.7)*	6 (17)	2.4 (4.0)

* The experimental results, obtained in isobutylene polymerization ($AlCl_3$ in chloroethane; T° is 243 K).

** The difference of calculated and experimental data for the values of $\bar{P}_n$ and $\bar{P}_w/\bar{P}_n$ is exlained by the choice of correlation k_p/k_m and when necessary can be easily eliminated.

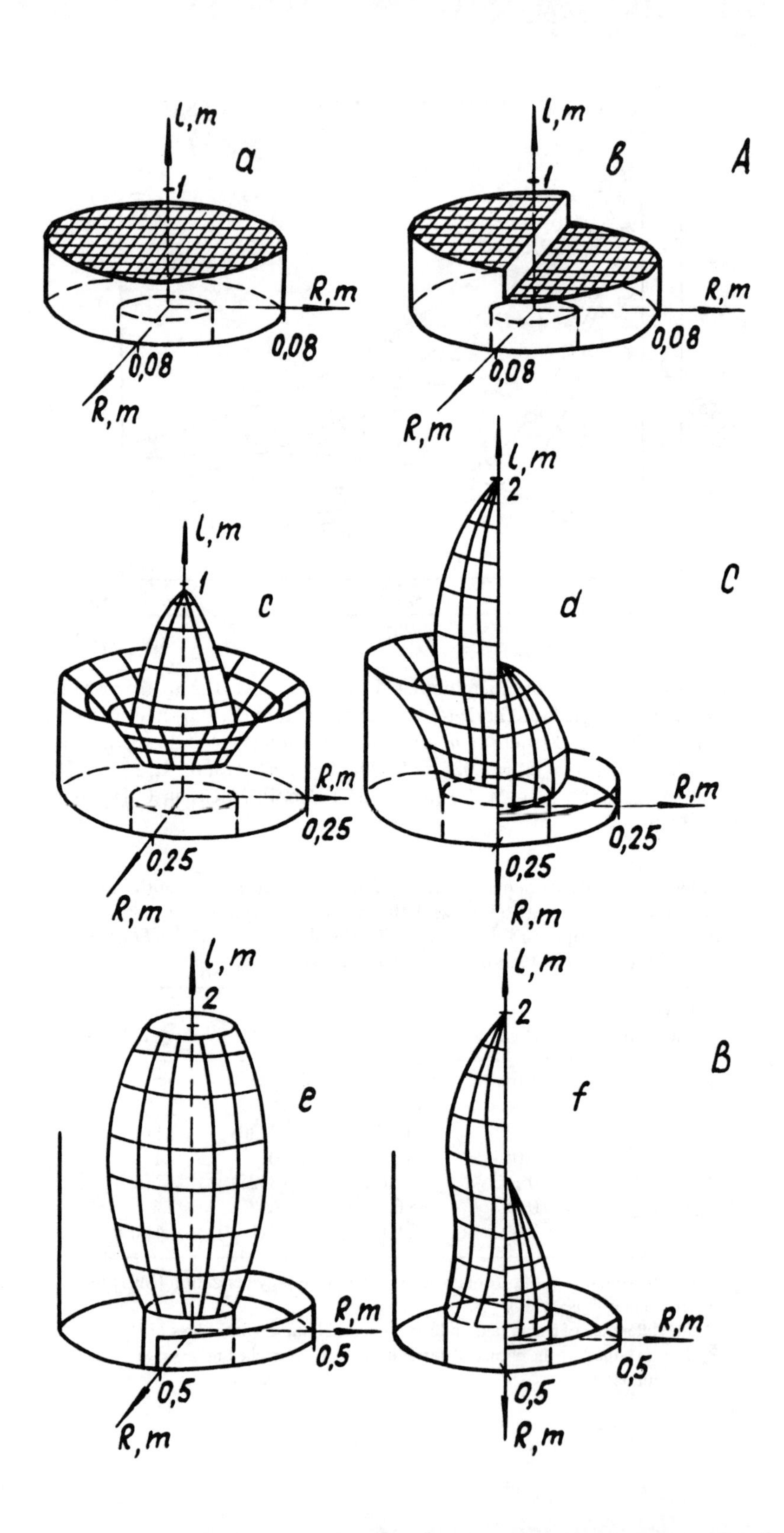

l, m
a
1
R, m
0,08
0,08
R, m
l, m
в
1
A
R, m
0,08
0,08
R, m
l, m
1
c
R, m
0,25
0,25
R, m
l, m
2
d
C
R, m
0,25
0,25
R, m
l, m
2
e
R, m
0,5
0,5
R, m
l, m
2
f
B
R, m
0,5
0,5
R, m

mixing (diffusion) of cocurrent flows and the termination of active centres. The reaction proceeds without reaching the reactor walls and the conversion is markedly reduced due to the monomer slipping between the "torch" boundary and the reactor wall.

Type B (the third macroscopic type) determines the intermediate process condition with the formation of "torch" having no the monomer slip zones. Types C and B are characterized by the gradients (fields) of temperature, active sites and the monomers alongs the axis X as well as along R (Fig. 15c, d), that considerably influences the uniformity of the polymer product formed (the MWD is broadened).

Thus, in modeling the liquid-phase electrophilic polymerization of isobutylene, one can prove the apparent effect of the reaction volume geometry, i.e., the transition from one macrokinetic reaction type to another (types A, B, C) on the molecular characteristics of the products formed (Table 3). When increasing the radius R, the MWD width ($\bar{P}_w/\bar{P}_n$) grows. The mean numerical MW of the polymer product simultaneously decreases. It is connected with the fact that the temperature in the volume is distributed in a relatively uniform way in the reaction zone of the small R, while the formation of the temperature gradient in the form of the "torch" along the reaction volume coordinates at the radii higher than a certain R_{cr} (critical) (R_{cr} is a certain value of R which stipulates the transition from the conditions of type A to those of type B) results in the MWD broadening due to the accumulation of the portion of the low-molecular fraction. At the same time one should bear in mind that the MWD of the polymer product broadens as the catalyst supply points move away along the axis X, which is the result of the temperature increase. The data in Table 3 show that the computation correctly reflects the effect of the reaction zone radius R on molecular characteristics of the polymer formed as well as on monomer conversion and is in accordance with the experimental results.

Thus, in liquid-phase fast polymerization, one can observe the phenomenon—the influence of the geometrical parameters of the reaction zone on the final product yield and its molecular characteristics. Unlike classical chemical processes, fast liquid-phase polymerization

Figure 15. Fields of temperatures (a, c, e) and of monomer and active centers concentration (b, d, f) in fast polymerization of monomer in flow for $R, m = 0.08$ m (a, b); 0.25 (c, d) and 0.5 (e, f) in maximum for the given $\bar{P}_n$ conversion, per cent by weight 99.3 (a, b); 65.0 (c, d); 32.0 (e, f) at $\Delta T^\circ = 20$ (a, b); 22 (c, d); 9 (e, f) ($[M]_0 = 1$ mol/l; $[A^*]_0 = 4.5 \cdot 10^{-3}$ mol/l; $D_T = 0.025$ m^2/s; $k_p = 10^5$ l/mol·s; $k_t = 20$ s^{-1}; $R/r = 2.5$; $T_0 = 300$ K). Macroscopic regimes of processes: Type A (a, b); Type B (e, f); Type C (c, d).

has the characteristic that the increase of the reaction zone volume and the supply of greater amount of the reaction mixture with the same linear rate results in the marked decrease of the conversion and considerable reduction of the reactor productivity.

2.2 THE CORRELATION BETWEEN THE KINETIC AND HYDRODYNAMIC CONSTANTS AND THE PARAMETER OF REACTION ZONE GEOMETRY

The transition from new conditions of the quasi-ideal displacement in turbulent flows (type A), which have no analogues, to the "torch" ones (type C and especially type B) is accompanied by the monomer conversion decrease and the deterioration of the MW of the polymer ($\bar{P}_n$ is reduced and the MWD is broadened). The critical radius R_{cr}, defining the transition of the reactor operation from the conditions of type B to the ones of type A, is determined by the ratio of the diffusion to the destruction of active sites A* (D_T and k_t). Then R_{cr} is expressed in the form:

$$R_{cr} = m(D_T/k_t)^{1/2} \tag{18}$$

where R_{cr} is a value of the reaction zone radius that secures the decrease of the polymer yield by 10% as compared with the one at $R \to 0$.

According to the results of the mathematical modeling, such ratio is valid (Fig. 16) at the coefficient of proportionality $m = 2 \pm 0.2$ [34, 35].

It should be noted that in changing the radius (at the constant linear rate of the flow), three different zones of the liquid flow may be singled out: the laminar (at small R), the transitional, and the turbulent (at great R) conditions. Each type of conditions has its own values of the average time of mixing ($\tau_{mix} \sim R^2/D$) (Fig. 17).

The coefficient of diffusion of the laminar type of conditions is very small: $D = 10^{-9}\, m^2/s$. As a consequence, the time of mixing is great, and it increases with R enlarging R^2. The transitional regime is characterized by the increase of the mixing effeciency (D) and decrease of τ_{mix}. For the turbulent conditions the value of D_T, which is calculated to be $D_T = 10^{-3} - 1\ \mathrm{m^2/s}$, is very great and it increases approximately according to the law $D_T \sim R$. Comparing τ_{mix} with the characteristic time of the chemical reaction τ_{ch}, one can see that the conditions of quasi-ideal displacement ($\tau_{mix} < \tau_{ch}$) are maintained in the intervals $0 < R < R_1$ and $R_2 < R < R_3$. The former ($0 < R < R_1$) corresponds to the capillary reactors (for $D = 10^{-9}\ \mathrm{m^2/s}$ and $\tau_{ch} \cong 1$ s, $R_1 \simeq 3.10^{-5}$ m) and presents no practical interest.

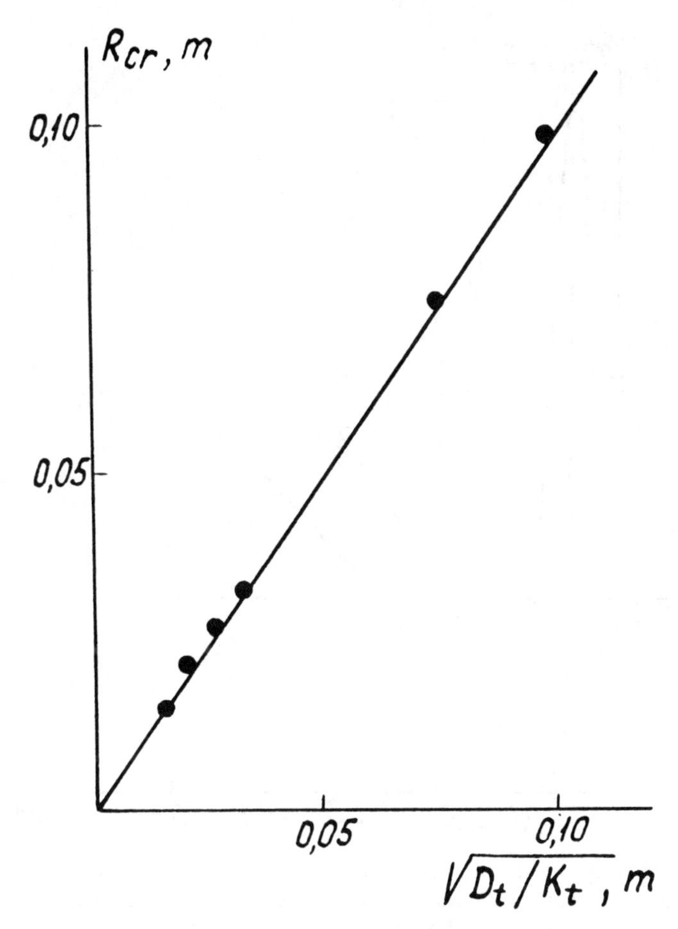

Figure 16. Relationship between R_{cr} and $(D_T/k_t)^{1/2}$.

Thus, there remains rather narrow (depending on τ_{ch} and V) interval of the turbulent flow $R_2 < R < R_3$, having very high productivities (Fig. 17). For instance, at $V = 1$ m/s the productivity is 20 m^3/h, and in the range of radii from $R_2 = 0.03$ m to $R_3 = 0.15$ m, the productivity of the tubular reactor varies from 20 to 500 m^3/h. The increase of the flow rate results in the expansion of the possible sphere of employing the reactor (R_2 is decreased while R_3 is increased), but the hydrodynamic resistance of the reactor may increase. Consequently, it is necessary to raise the pressure, which may cause undesirable effects (the increase of

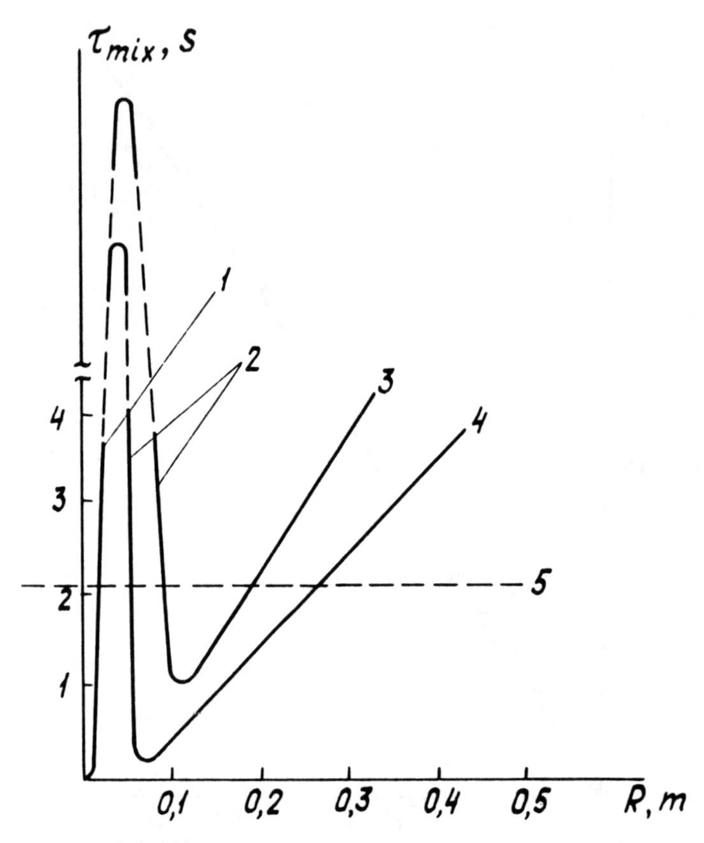

Figure 17. Dependence of time of mixing $\tau_m = R^2/D_T$ on the flow radius R. Conditions: 1 – laminar; 2–transitional; 3,4 – turbulent; linear rates of the flow V m/s: 3 – 2.5; 4 – 5.0. Break line (5) correspond of $\tau = 1/k_t$.

the boiling temperature of the reaction mass, toughened requirements to the apparatus design, greater loads on pumps, etc.)

The determination of the length 1 of the reaction front spreading along the axis x is the important aspect of the problem connected with the proceeding of very fast processes of polymerization. Since the reaction zone length is related to the rate of polymerization through the linear rate of the concurrent flows, there is a correlation between the effective time of the reaction mass staying in the reaction zone (I/V) and the effective time of the chemical reaction which constitute a combination of values $1/k_t$ and $1/k_p\ [A^*]_0$.

The yield of the polymer, expressed in the analytical form for the reactor of ideal displacement, is defined by the ratio:

$$\beta = 1 - \exp\left\{ -\frac{k_p[A^*]_0}{k_t}\left[1 - \exp\left(-k_t\frac{1}{V}\right)\right]\right\} \tag{19}$$

If we introduce the parameter $(1/V)_{ef}$, which is the effective time of the stay defining such reaction zone length that secures the polymer yield of 90% of the maximum one at $1 \to \infty$, then:

$$(l/V)_{ef} = -\frac{1}{k_t}\ln\left\{1 + \frac{k_t}{k_p[\mathrm{A}^*]}\ln\left[0,1 + 0,9\exp\left(-\frac{k_p[\mathrm{A}^*]_0}{k_t}\right)\right]\right\} \tag{20}$$

In this case the dependence $(l/V)_{ef}$ on $(1/k_p[A^*]_0)$ has the form given in Fig. 18.

The two extreme cases are of importance. At the great catalyst concentrations $k_t/k_p[A^*]_0 \ll 1$ the polymer yield approaches the quantitative one (100% wt). Then $(l/V)_{ef} \cong 2.3/k_p\ [A^*]_0$. At small catalyst concentration $k_t/k_p\ [A^*]_0 \geqslant 1$ and $(l/V)_{ef} \cong 2.3/k_t$.

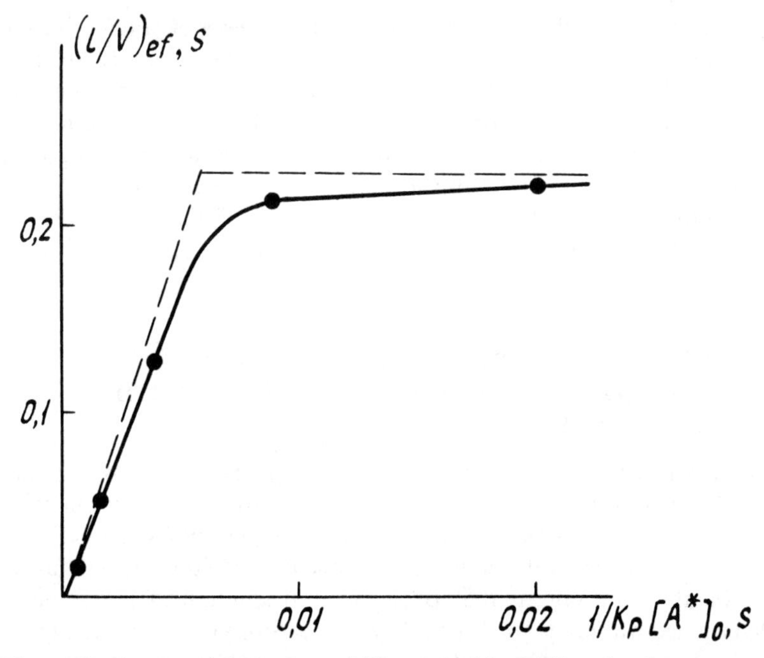

Figure 18. Computed dependence $(l/V)_{ef}$, s on $1/k_p\ [A^*]_0$, s for the reactor of ideal (continuous line) and quasi-ideal (○) displacement: $R = 0.08$ m; $V = 2.5$ m/s.

The computation of very fast processes of polymerization in the reaction zone, taking into account the longitudinal diffusion (but under the condition that $R < (D_T/k_t)^{1/2}$), has shown that the curves of dependence of $(l/V)_{ef}$ on $1/k_p$ $[A^*]_0$ at different (in the wide range) values of $[A^*]$ are close to those calculated by equation (20) (Fig. 18).

As is known, the determination of the elementary rate constants of reaction for fast processes of polymerization is an extremely difficult task. But results given above allow us to determine the main elementary constants experimentally. Studying the dependence of the polymer yield on the reaction zone length and/or on the flow rate at different catalyst concentrations and under the condition that R is less than R_{cr}, one can easily evaluate the rate constants k_p and k_t if either both parts of the correlation $k_t/k_P\,[A^*]_0 < 1$ or at least one of them is employed (when it is impossible to cover both parts of the correlation) [20].

The suggested method of computing the kinetic constants made it possible to estimate k_p and k_t for the cationic liquid-phase polymerization of isobutylene ($AlCl_3$ – $3.7 \cdot 10^{-3}$ mol/l, isobutylene $= 3.5$ mol/l), which at 243 K appeared to be: $k_p = 1.0 \cdot 10^6$ l/mol·s (that corresponds to the data given in [18, 19]) and $k_t = 17.5 \pm 5\,s^{-1}$ (it was evaluated for the first time).

Thus, the kinetic parameters of the fast processes of polymerization define the geometrical size of the reaction zone. There is a correlation between the elementary process constants k_p and k_d and the geometrical parameters of the reaction zone (R, 1) and V. Herewith, quite new methods of the process control (unlike the standard ones being only kinetic or thermodynamic) appear to determine the depth of the monomer conversion and the mass molecular characteristics of the polymer product formed, in particular, the forced change of the reaction zone radius R.

2.3 THE LINEAR FLOW RATE EFFECT ON THE MASS MOLECULAR CHARACTERISTICS OF THE PRODUCT FORMED

The coefficient of turbulent diffusion D_T may be varied over rather a wide range at the expense of preturbulization, by changing the method of mixing and the direction and rate of the reagent flows [36–38], etc. Fig. 19 represents the dependence of the coefficient of turbulent diffusion D_T on the flow rate at the concurrent introduction of the reagents determined within the "q-ε" model by Navie–Stoks [39] considering the real values of the viscosity of the solution, its temperature, the rate of heat evolution etc. Obviously, depending on the movement of the

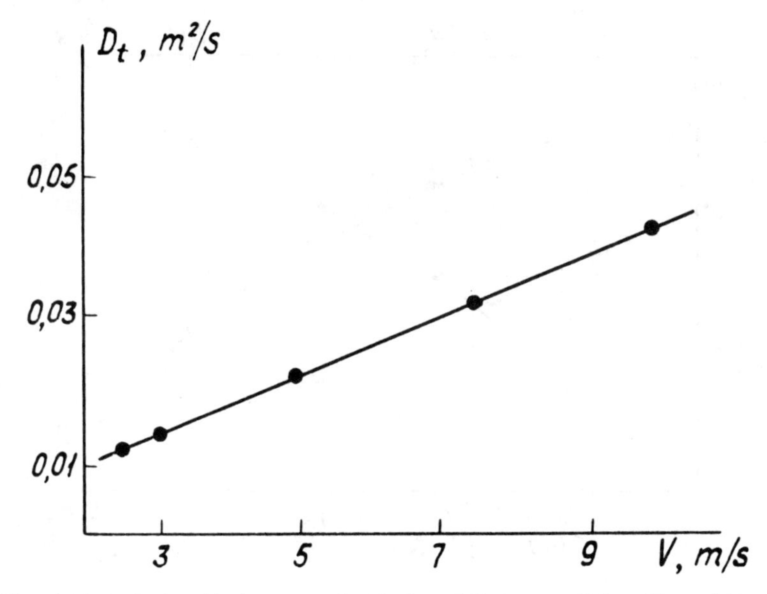

Figure 19. Relationship between of turbulent diffusion coefficient (D_T, m^2/s) on the flow rate of monomer and catalyst (V, m/s).

reacting flows (the monomer, the catalyst), the coefficient of turbulent diffusion increases in an almost linear way [29].

Substituting the numerical values of D_T obtained from the equations of Navie–Stoks (Fig. 19) in the system of equations (6)–(8) that describe the change of the monomer and active centers concentrations as well as the temperature in the reaction zone for a tubular reactor with the cocurrent reagent supply [31], it is easy to evaluate the effect of D_T, and of V, on the basic parameters of very fast polymerization processes.

For the polymerization proceeding in reactors with a radius greater than R_{cr}, the monomer conversion does not usually reach 100% wt [34, 35]. The increase of the coefficient of turbulent diffusion D_T in several times, including the one due to the growth of the flow rates, leads to a marked increase of the monomer conversion depth (at $R > R_{cr}$), despite the reduction of the time that monomer stays in the reaction zone (the length of the reaction zone 1 is constant) (Fig. 20, Curve 1).

If the reaction zone is limited by a single time that the monomer stays, for instance $\tau = 0.2$s, (Fig. 20, Curve 2), other things being equal, for all values of the flow rate the monomer conversion may more than triple, with the flow rate growing from 2.5 to 10 m/s is the fast polymerization processes. Simultaneously, and this is important, the increase of

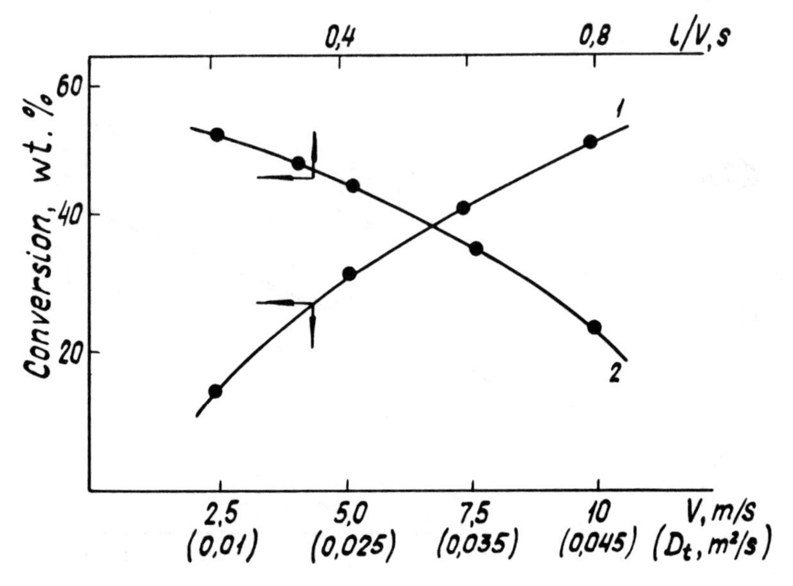

Figure 20. Dependence of polymer yield on the reaction time l/V, s-(1 – for $l = 2$ m) and flow rate (2 – for $1/V = 0.2$ s) ($[M]_0 = 2$ mol/l; $[A_0^*] = 0.0045$ mol/l; $R = 0.25$ m; $T_0 = 300$ K).

the monomer conversion depth (the increase of V as well as of D_T) has given rise to the change of massmolecular characteristics of the polymer product. The increase of the flow rate results in that of the average-numerical MW ($\bar{P}_n$), the MWD of the product being narrowed at the same time (Fig. 21). The temperature changes by R for different values of the linear rates of the reagent flow V(the turbulent diffusion coefficient D_T) are given in Fig. 22. One can see that at the increase of V and, correspondingly, the growth of D_T, the leveling of the temperature maximum in the reaction volume takes place, in spite of the fact that the total yield of the polymer increases. The dissipation of the heat energy results in the growth of the mean MW and narrowing of the MWD of the product.

It should be particularly noted that despite the value of the coefficient of turbulent diffusion in the regions with radii more than R_{cr}, the external heat removal α(α-Nu/2R) exerts practically no influence upon product MW and MWD. This is caused by the fact that the diffusion is not sufficient to reach the thermostating wall. The active centers terminate without reaching it, i.e., the process runs in the local zone, the size of which is determined by the ratio of the two competing processes—the

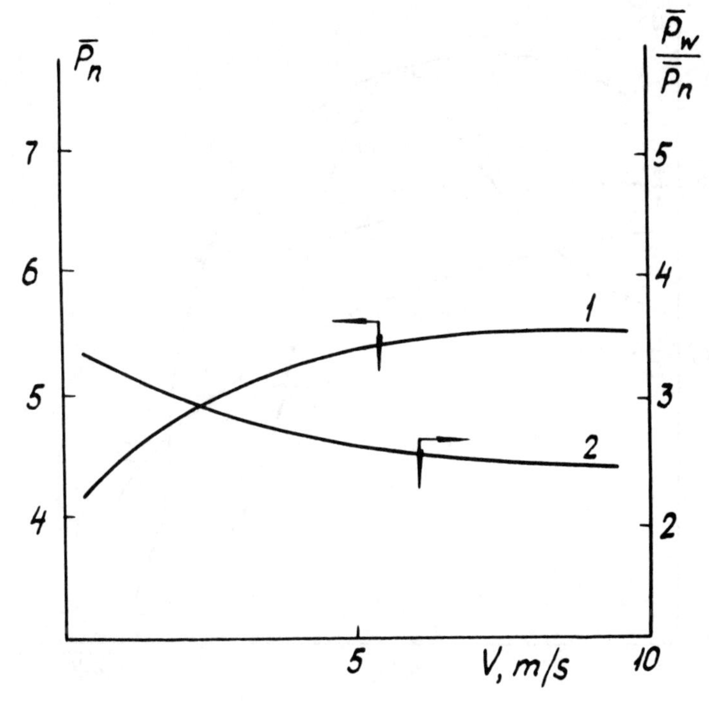

Figure 21. Dependence of $\bar{P}_n$ (1) and $\bar{P}_w/\bar{P}_n$ (2) on flow rate V, m/s for $[M]_0 = 2$ mol/l; $[A^*]_0 = 0.0045$ mol/l; $R = 0.25$ m; $T_0 = 300$ K; $\tau_{ch} = l/V = 0.2$ s.

active centers diffusion and their determination. The mean molecular mass growth and narrowing of the MWD of the polymer formed are due in this case to better distribution of the reacting components but not to heat removal in the volume; that is, they result from dissipation that creates more uniform distribution of the temperature fronts, to the point that there are no distinct high-temperature zones.

The condition of the low sensitivity of the WM and MWD to the reaction zone temperatures (more accurately, to the temperature gradient) that determines the quasi-isothermal regime of the process is expressed by the ratio:

$$\Delta T \ll \frac{R \cdot T^2}{E_m - E_p}, \tag{21}$$

where T is the average temperature, and ΔT is the difference of temperatures in the reaction zone.

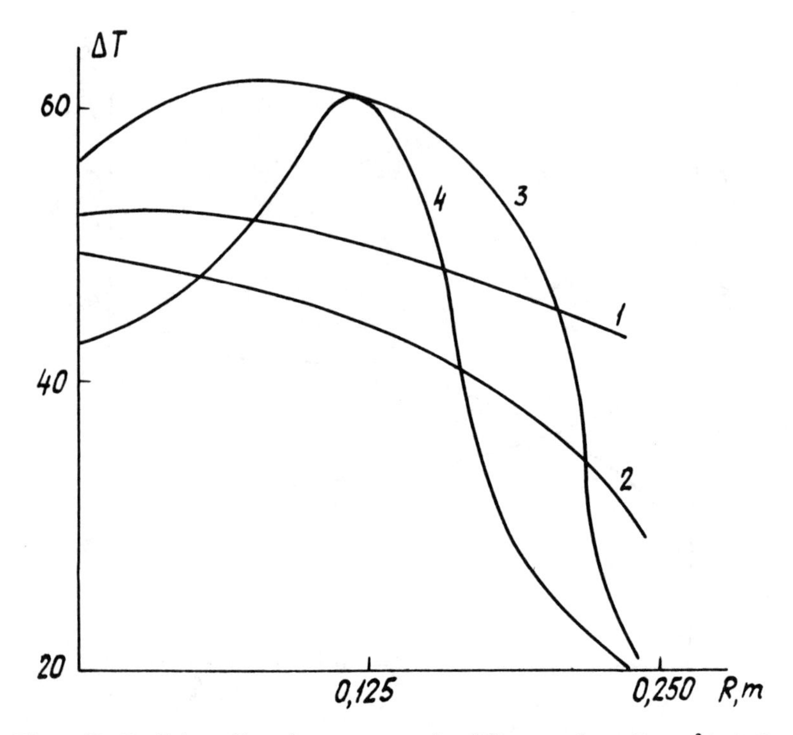

Figure 22. Radial profiles of temperature for different values: D_T, m^2/s: 1, 2 – 0.0045 ($V = 10$ m/s); 3, 4 – 0.01 ($V = 2.5$ m/s) depth of monomer conversion (q per cent by weight): 1, 3–90; 2, 4–60.

The temperature gradient ΔT in the reaction zone is determimed by the correlation of rates of the process of heat evolution in polymerization W_p and its distributin on account of the turbulent transition W_t:

$$W_T = \frac{\lambda_T \Delta T}{\delta} = \frac{c_p \cdot \rho \cdot D_T \cdot k_t \Delta T}{V}; \qquad W_p = \rho \cdot V \cdot q \cdot \Delta M \tag{22}$$

where $\delta = V/k_t$ is the reaction zone length at $k_p[\text{A}^*]_0 \ll k_t$. The combination of Eqs. (21) and (22) gives the following:

$$\frac{k_t \cdot \lambda_T \cdot RT^2}{\rho V^2 q \Delta M (E_m - E_p)} \gg 1 \tag{23}$$

Thus, it is necessary to follow two criteria, ratios (18) and (23), in order to maintain the conditions of the formation of quasi-isothermal regime in the reactors of displacement in turbulent flows.

The characteristic size of spreading zone (averaging) of the reaction temperature is proportional to $\lambda_T^{1/2} \sim D_T^{1/2} \sim V^{1/2}$, since in the turbulent conditions $D_T \sim \lambda_T \sim V$, the temperature shift along the reactor axis and correspondingly the increase of the reaction zone length are proportional to V(linear dependence). Therefore, the flow rate increase contradicts the ratio (23) and conditions at $V \rightarrow \infty$ approach those of the ideal displacement (in the classical variant) with the rather wide product MWD.

The flow rate (V) reduction leads to the transition from turbulent flow conditions to laminar ones, with the criterion (23) failing to be satisfied due to the sharp reduction of λ_T. This also means the breaking of the conditions of quasi-isothermal regime. In other words, there are a limited range of flow speeds for realizing the quasi-isothermal conditions when the tubular reactor is working.

Thus, by changing the flow rate one can essentially increase the polymer product yield with the simultaneous growth of the MW and the improvement of its quality (the MWD narrowing); i.e., one can effectively influence the process of polymerization and molecular characteristics of the products formed in changing the process hydrodynamics.

This is related to the change of the profile of temperature fields along the reactor coordinates in the reaction zone. In spite of the reduction of the time that the reagent being polymerized is in contact with the thermostating wall of the reactor ($\tau_{mix} = 1/V$), the increase of V will lead to a sharp increase in the effectiveness of inner heat removal, which causes the increase of the MW and the narrowing of the MWD at the quantitative yield of the polymer Product. This is simply impossible under classical kinetic conditions.

2.4 THE EFFECT OF THE METHOD OF CATALYST INTRODUCTION ON THE MASS-MOLECULAR CHARACTERISTICS OF THE POLYMER FORMED

In fast polymerization in the flow, at small radii of the reaction zone ($R < R_{cr}$), the front of the reaction propagation is planar, as has been shown above, and the temperature gradient by the flow radius is lacking (Fig. 15). This results in the formation of the polymer product with the MWD close to the exponential one. At the same time at great radii of the reaction zone ($R > R_{cr}$), the formation of the "torch" conditions takes place, characterized by the gradient (fields) of temperature along the reaction zone coordinates. As a consequence, the polymer with the wide MWD is produced, the present thermostating system being inefficient [23]. However, the change of the system of the reagent supply into the reaction zone markedly influences the character of fast polymerization.

The effect of the supply method of the monomer M and the catalyst A* in the reaction zone in the flow isobutylene polymerization has been studied within the unified mathematical model [37, 38]. Two models of the reagent supply have been chosen, namely: a) with the central single point introduction of the catalyst; b) with the catalyst supply along the outer surface of the pipe (Fig. 23).

The equations (6)–(8) were used for calculation with border conditions:

Model I

$$T(0,r) = T_0$$

$$M(-d,r) = \begin{cases} M_0, & R > r > r_0 \\ 0, & r < r_0 \end{cases}$$

$$A^*(-d,r) = \begin{cases} 0, & R > r > r_0 \\ A_0^*, & r < r_0 \end{cases}$$

$$k_t(x,r) = \begin{cases} k_t, & x \geqslant 0 \\ 0, & x < 0 \end{cases}$$

Model II

$$T(0,r) = T_0$$

$$M(-d,r) = \begin{cases} 0, & R > r > r_0 \\ M_0 & r < r_0 \end{cases}$$

$$A^*(-d,r) = \begin{cases} A_0^* & R > r > r_0 \\ 0, & r < r_0 \end{cases}$$

$$k_t(x,r) = \begin{cases} k_t, & x \geqslant 0 \\ 0, & x < 0 \end{cases}$$

$$\frac{\partial M(x,r)}{\partial r} = \frac{\partial A^*(x,R)}{\partial r} = \frac{\partial A^*(x,0)}{\partial r} = \frac{\partial M(x,0)}{\partial r} = \frac{\partial T(x,0)}{\partial r} = 0$$

$$\frac{\partial T(x,R)}{\partial r} = \alpha\{T(x,R) - T_1\}$$

Here x and r are the coordinates along the length and the radius of the reaction zone; T, T_0, T_1 are the temperatures of the reaction, reagents and external thermostating respectively; α–the coefficient of heat transfer through the wall ($\alpha = \mathrm{Nu}/2R$), $k_t = k_t^0 \exp(E_t/RT)$ is the chain termination rate constant.

The numerical solution of the system of equations (6)–(8) was carried out with the help of the implicit difference scheme and the MW and MWD were calculated according to [28].

At small radii of the reaction zone (Fig. 24) without outer heat removal, the method of the reagent supply exerts no influence on $\bar{P}_n$ and $\bar{P}_w/\bar{P}_n$ of the product; with the increase of the outer heat removal efficiency, the MW and MWD of the product differ, depending on the method of the reagent supply. Thus, if in approaching model I, $\bar{P}_n$ increases and $\bar{P}_w/\bar{P}_n$ linearly decreases approximating the exponential

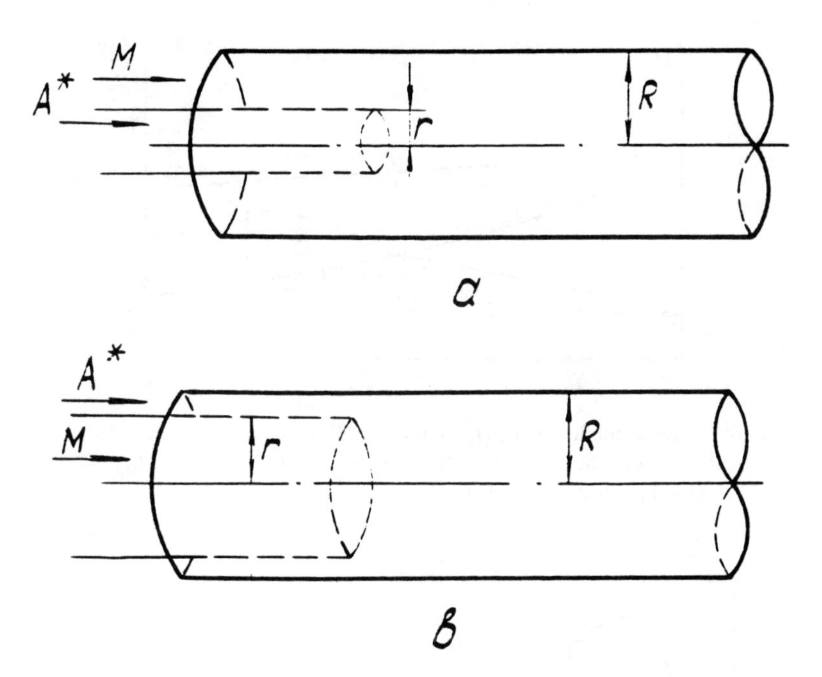

Figure 23. Model I (a) and model II (b) of the monomer and catalyst supply in a tubular reactor.

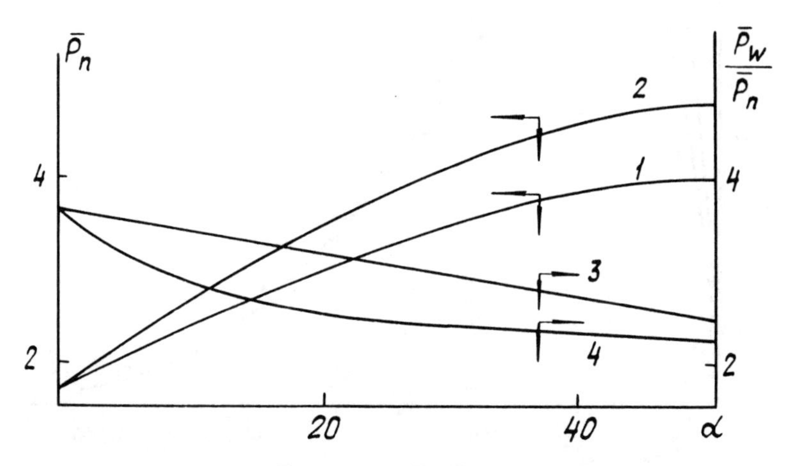

Figure 24. Dependence of $\bar{P}_n$ (1, 2) and $\bar{P}_W/\bar{P}_n$ (3, 4) on the efficiency of heat removal (α) in the cocurrent (1, 3) and radial (2, 4) ways of the catalyst supply ($[M]_0 = 2$ mol/l; $[A^*]_0 = 0.0045$ mol/l; $R = 0.066$ m: $T_0 = 300$ K).

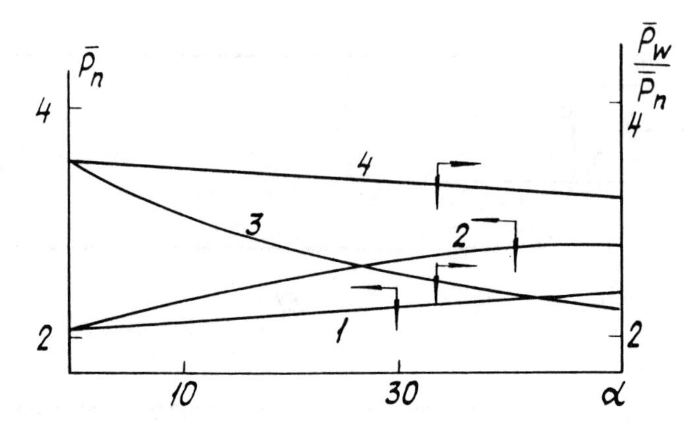

Figure 25. Dependence of $\bar{P}_n$ (1, 2) and $\bar{P}_W/\bar{P}_n$ (3, 4) on the efficiency of heat removal (α) for model I (1, 4) and model II (2, 3) ($[M]_0 = 2$ mol/l; $[A^*]_0 = 0.0045$ mol/l; $R = 0.264$ m; $T_0 = 300$ K).

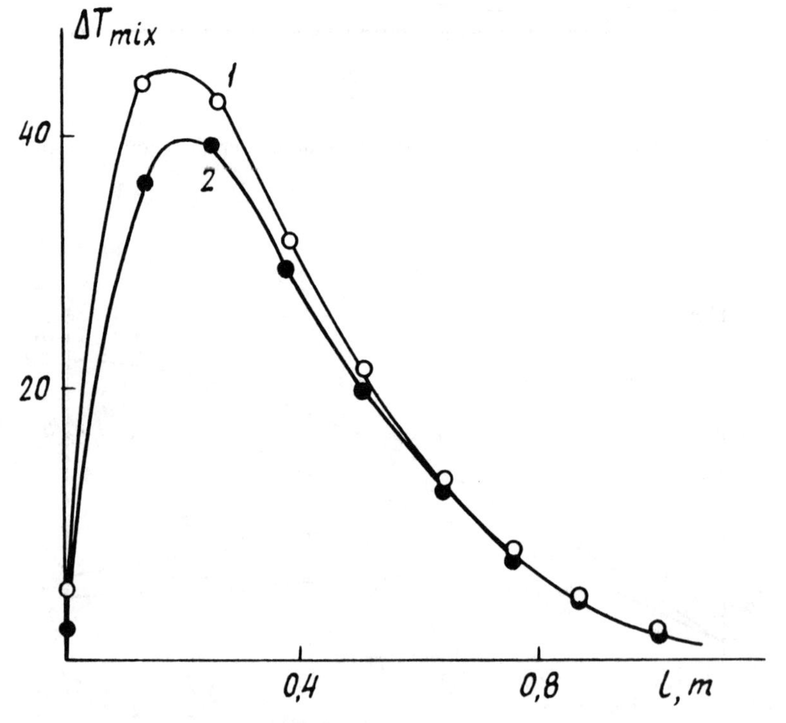

Figure 26. Dependence of the mean-in-the-cross-section change temperature (ΔT^0_{mid}) on the reactor length (l, m) for model I (1) and model II (2) ($[M]_0 = 2$ mol/l; $[A^*]_0 = 0.0045$ mol/l; $R = 0.0066$ m, $T_0 = 300$ K).

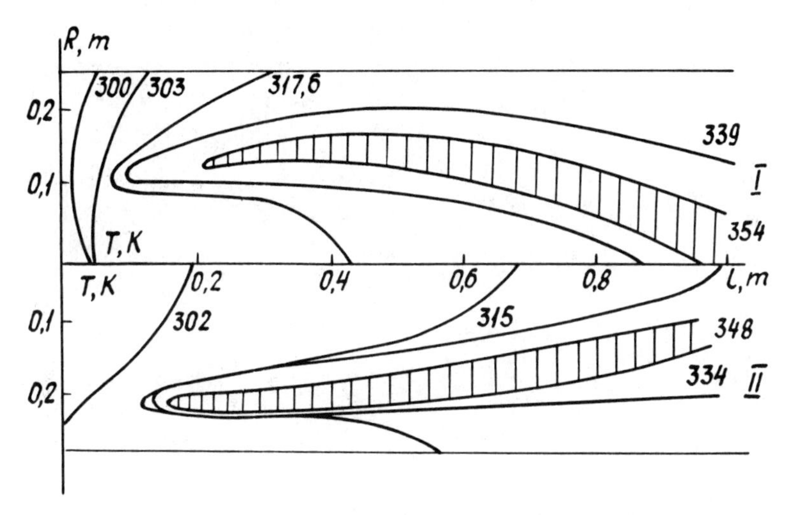

Figure 27. Fields of temperatures in polymerization of isobutylene with external heat removal for models I and II ($R = 0.264\,\text{m}$). The region of the maximum temperatures is crosshatched. The figures on curves are temperatures, K.

1 at $\alpha = 50$, in model II, the MWD approaches 2 already at $\alpha = 20$. It is of interest that $\bar{P}_n$ for model II is always higher than for model I under similar conditions.

With the reaction zone radius increase, the effect of the change of polymer molecular characteristics becomes more tangible (Fig. 25). In this case for model *I*, $-\bar{P}_n$ is slightly changed with the increase of α and $\bar{P}_w/\bar{P}_n$ does not practically depend on α. At the same time, when model II is used, $\bar{P}_w/\bar{P}_n$ considerably changes and approaches the exponential 1 simultaneously with growth of $\bar{P}_n$.

The established effect is related to the character of the distribution of the temperature front in the reaction volume. At small radii, i.e., when the planar reaction front is being formed, the thermostating effect is due to the reduction of the reaction temperature at the initial stage of the process, and it changes uniformly along the reaction zone length for both models (Fig. 26). At great radii, the character of the reaction front changes as well (Fig. 27). Thus, model II shows the formation of the maximum temperature zone near the thermostating wall, which considerably contributes to heat removal and levels the temperature field. As a consequence, the MWD is narrowed.

3 REGULATION OF THE HEATING REGIME IN FAST POLYMERIZATION PROCESSES

3.1 THE HEATING REGIME IN POLYMERIZATION WITHOUT HEAT REMOVAL

The most important factor that determines the efficiency of very fast polymerization is the temperature field along the reaction zone coordinates. The form of that field depends on the geometry of the reactor-polymerizer, the monomer and the catalyst concentrations, the flow rates and turbulence of the reaction mixture and the reagents, and the methods by which components are mixed. With this in mind, we should pay special attention to the formation of the reaction temperature front and to the control over the thermal regime of the reaction [39–42].

In fast polymerization when $R < R_{cr}$ [31–33] (Eq. (18), Fig. 16), the intense longitudinal and latitudinal mixing averages the temperature in the reaction zone so that the MWD and the mean MW turn out to be close to those characteristic of isothermal conditions at temperatures corresponding to the adiabatic heating of the reaction medium. However, the possibility of polymerization under adiabatic conditions is usually limited by the fact that at high temperatures the MW considerably decreases and other by-processes begin, in particular, those of degradation, cross-linking of macromolecules, etc.

In cationic polymerization (of isobutylene, in particular), the temperature at which the process may be conducted is mainly limited by the chain transfer to monomer that leads at 303 K to the formation of oligomers with the degree of polymerization of several units. Therefore, it is necessary to know the possible ways of conducting isobutylene polymerization under adiabatic conditions and to determine the boundaries of probable heating of the initial reaction mixture as well as the excess heat removal on account of external thermostating or inner heat removal.

The possibility of cooling the initial mixture, in the case of process realization when the boiling temperature of the reaction mass is not reached, is restricted by the boiling temperature of the cooling agent and of liquid ethylene in particular ($T_{boil} = 183$ K). The temperature of the mixture will grow proportionally to the polymer yield increase until the solvent starts boiling according to the correlation:

$$\Delta M = \frac{\bar{c}_p}{q_{11}}(T - T_0) = 0.00175(T - T_0) \tag{24}$$

Here T_0 is the feed temperature at the reactor inlet; T is the temperature of the polymerizate; $\bar{c}_p$ is the average thermal capacity (1.68 J/g·deg); q_{11} is the thermal effect of isobutylene polymerization that is equal to 970 J/g.

In case of using the range of temperature variation of 363 K during the heating of the reaction mass using ethylene as the cooling agent, nearly 2.1 mol/l of the dissolved isobutylene can be polymerized in the given concrete conditions. If the cooling action of the monomer solution is lower for instance, if ammonia is used as the cooling agent with an operating temperature of 243 K—one can obtain no more than 10%wt of the polymer during the reaction mass heating. Figure 28 represents the

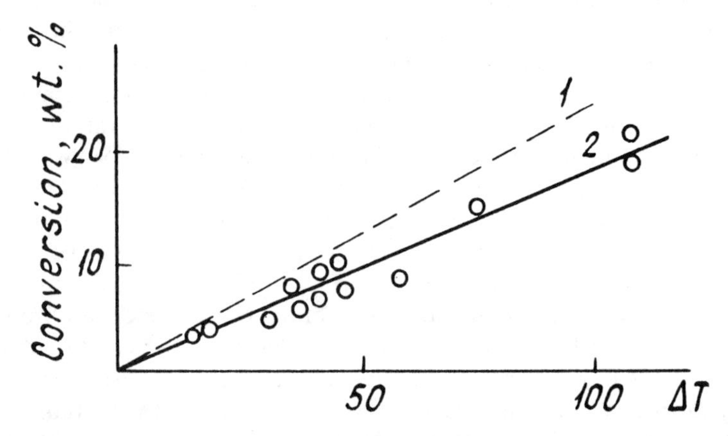

Figure 28. Dependence of the temperature difference (ΔT) at the reaction mixture in flow and out flow from the reactor on the polymer yield: 1 – theoretical dependence; 2 – experimental dependence; $[M]_0 = 3.2$ mol/l; $[A^*]_0 = 0.0045$ mol/l ($AlCl_3$ in C_2H_5Cl); $T_0 = 213$ K.

dependence of the temperature difference of the reaction mixture at the reactor inlet and outlet on the polymer yield that allows us to determine either the average temperature of the reaction in each zone or the amount of the polymer formed in these zones if one of the parameters is known.

3.2 THE HEATING REGIME OF FAST POLYMERIZATION IN THE CONDITIONS OF INNER HEAT REMOVAL DUE TO THE REAGENT BOILING

One of the ways to remove the chemical reaction heat is keeping the given temperature constant in the reaction volume by the boiling of either the solvent or the monomer.

Boiling takes place in a certain finite, rather narrow temperature range, with the boiling temperature depending in a complex way on the amount of the monomer polymerized, on the MW of the polymer formed etc. For a proper account of this dependence it is necessary to know the variation of external heat of vaporization due to the reaction medium content.

In the first approximation, we may consider that boiling occurs at the constant temperature, especially as in a number of processes the external heat of vaporization is quite sufficient up to the end of polymerization (the initial monomer concentration $[M]_0 = 2.1–3.5\,\mathrm{mol/l}$ in a chlorinated solvent). The careful analysis of the thermal balance is given in [40]. The reaction temperature increases in all zones until it reaches T_{boil}. Then the process proceeds at a practically constant temperature until the boiling stops, after all the solvent and/or a certain portion of the monomer have boiled out. During the transition of all the boiling liquid into vapor the temperature of system again starts increasing greatly due to the polymerization of either liquid or gaseous monomer (the monomer does not boil in the system).
Then

$$T_i = \begin{cases} T_0 + \alpha \cdot \sum\limits_{k=1}^{i} \Delta M_i & \text{at} \quad T_0 + \alpha \cdot \sum\limits_{k=1}^{i} \Delta M_i \ll T_{\mathrm{boil}} \\ T_{\mathrm{boil}} & \text{at} \quad T_0 + \alpha \cdot \sum\limits_{k=1}^{i} \Delta M_i > T_{\mathrm{boil}} \\ & \text{and} \quad \sum\limits_{k=1}^{i} \Delta M_i < \Delta M^* \\ T_{\mathrm{boil}} + \xi\left(\sum\limits_{k=1}^{i} \Delta M_i - \Delta M^*\right) & \text{at} \quad T_0 + \alpha \cdot \sum\limits_{k=1}^{i} \Delta M_i > T_{\mathrm{boil}} \\ & \text{and} \quad \sum\limits_{k=1}^{i} \Delta M_i > \Delta M^* \end{cases}$$

Here $M^* = (\bar{c}_p(T_k - T_0) + \chi_{\text{boil}}S)/Q_{11}$ is the monomer conversion value when boiling ends and $\alpha = \alpha' = Q_{11}\bar{c}_p$, if the solvent is boiling.

The equation of thermal balance at the moment of transition of the monomer that has not been polymerized into the gas phase has the following form:

$$Q_{11} \cdot \Delta M^* = \bar{c}_p(T_{\text{boil}} - T_0) + \chi_s S_0 + \chi_m(M'_0 - \Delta M^*) \tag{25}$$

Hence,

$$\Delta M^* = \frac{\bar{c}_p(T_{\text{boil}} - T_0) + \chi_s S_0 + \chi_m M_0}{\chi_i + Q_{11}} \tag{26}$$

where $S_0 = (1 - M_0)$, M_0 are the initial weight proportions of the solvent and monomer in the mixture; S_0 and M_0 the amounts of the solvent and the monomer in the initial mixture, χ_s and χ_m are heats of vaporization of the solvent and the monomer respectively.

The dependences of ΔM on $[M]_0$ according to equation (26) for different cases and the real system are given in Fig. 29. At $M_0 \leqslant [\bar{c}_p(T_{\text{boil}} - T_0) + \chi_s]/(Q_{11} + \chi_s)$, all the monomer can be polymerized on account of heating the initial mixture and boiling the solvent

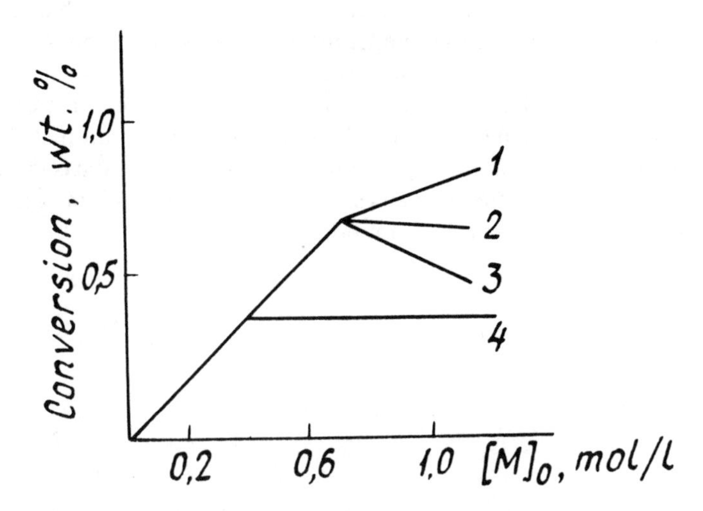

Figure 29. Computed dependence of the polymer yield on the mass portion of the monomer: $\chi_m > \chi_s$ (1); $\chi_m = \chi_s$ (2); $\chi_m < \chi_s$ (3); experimental dates for the real system of isobutane-isobutylene χ_m and χ_s J/deg are the heats of vaporization of the monomer and solvent respectively.

($\Delta M = M_0$). At $M_0 > [\bar{c}_p(T_{\text{boil}} - T_0) + \chi_s]/(Q_{11} + \chi_s)$, a part of the monomer evaporates and becomes gas. The more that evaporates, the less is χ_m. If external heats of vaporization of the monomer and the solvent are the same, we may obtain equal quantities of the polymer due to the monomer evaporation and to that of the solvent.

For instance, temperatures and heats of boiling for isobutylene χ_m and isobutane χ_s make up to 266 K and 400 J/g and 261.3 K and 367 J/g, respectively. In obtaining low-molecular polyisobutylenes on the basis of the isobutane-isobutylene fraction with the content of the main components about 50 $\pm$ 5 %wt, the isobutane boiling may completely remove the evolving heat in the monomer conversion by 35%wt (g/g of the solvent) or 70%wt of the initial monomer.

The quantitative criterion of the constant temperature in analyzing the MW and the MWD of the polymer is the ratio:

$$T''_{\text{boil}} - T'_{\text{boil}} \ll \frac{RT^2}{E_m - E_p} \tag{27}$$

where E_m and E_p are effective activation energies of the chain transfer on monomer and chain propagation.

When the temperature depends linearly on the quantity of the polymer formed:

$$\Delta T = \alpha_{\text{boil}} \Delta M \quad \text{at } T'_{\text{boil}} < T < T''_{\text{boil}}$$

we have if only the solvent is boiling

$$\alpha_{\text{boil}} = \alpha^s_{\text{boil}} = (T''_{\text{boil}} - T'_{\text{boil}})/\chi_s S Q_{11} = \alpha_2 \tag{28}$$

(Figure 30) and

$$\alpha^{s,m}_{\text{boil}} = \frac{T''_{\text{boil}} - T'_{\text{boil}}}{\Delta M^* - \Delta M} = \frac{(T''_{\text{boil}} - T'_{\text{boil}}) \cdot Q \cdot_{11} (Q_{11} - \chi_m)}{Q_{11}(\chi_s S_0 + \chi_m M_0) - \chi_m \bar{c}_p (T'_{\text{boil}} - T_0)} \tag{29}$$

if both the solvent and the monomer are boiling.

The validity criterion of such a linear model in analyzing the polymer MW and MWD in the first approximation appears to be correlation (21).

In case the gaseous monomer remains in the system after the boiling stops, the system will continue to heat up. The above relation slightly changes the average heat capacity of the mixture "gas-liquid" and the polymerization heat essentially increases $Q_{11} = Q_1 + \chi_m$. Accordingly,

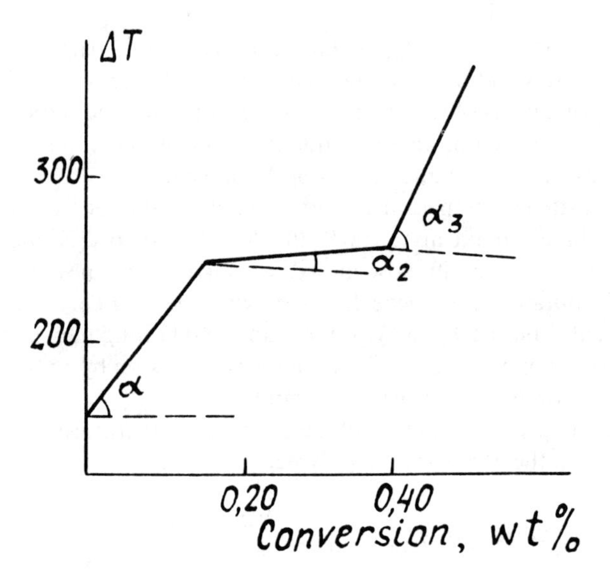

Figure 30. Relationship between of the temperature differences ΔT° and the polymer yield (it is diagram).

the slope of the curve of the dependence of ΔT on ΔM in this region will be equal to:

$$\alpha_3 \simeq Q_{11}/\bar{c}_p \quad (T > T''_{\text{boil}}, \text{ the monomer in liquid phase})$$

$$\alpha_3 \simeq \frac{Q_{11} + \chi_m}{\bar{c}_p} \quad (T > T''_{\text{boil}}, \text{ the monomer passed into gas phase}).$$

Here $\bar{C}_p$ is the average heat capacity of the liquid-gas mixture.

Thus, in general, inner heat removal on account of the boiling of the components of the reaction mixture proves to be an effective way of thermostating chemical reactions. The restriction of the reaction mixture temperature in liquid-phase fast polymerization due to the boiling of the part of the monomer or the solvent affects the reaction run in different ways depending on the reaction zone radius R [40]. In the region of small radii ($R < R_{cr}$), when the planar front of the reaction is formed and the temperature flow in the reaction zone is relatively uniform along the reaction zone radius (Chapter 2) [24–28], the process and the molecular mass characteristics of the polymer formed within a certain temperature range stop depending on the initial temperature of the feed (Fig. 31).

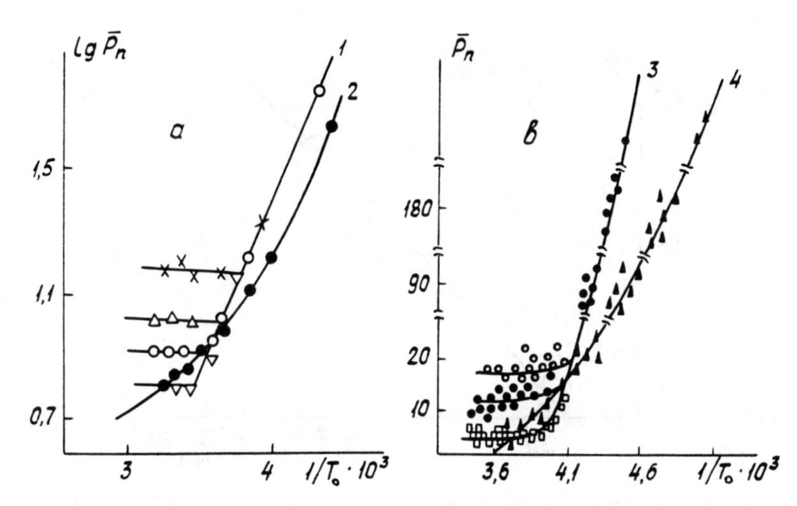

Figure 31. Dependence of lg $\bar{P}_n$ on $1/T_0$ for the different radii of the reaction zone of tubular turbulent reactor: a) computed curves at R, m: 1 – 0.08; 2 – 0.5 at $[M]_0 = 0.5$ mol/l; $[A^*]_0 = 0.0045$ mol/l; $T_{boil} = 297$ K; $\alpha = 0$; 2 – experimental dates at R, m: 3 – 0.025; 4 – 0.5 at $[M]_0 = 3.2$ mol/l isobutylene in isobutane; $[A^*]_0 = 1 \cdot 10^{-2}$ mol/l ($AlCl_3$ in C_2H_5Cl); $T_{boil} = 305$ K.

The monomer polymerization undergoes conditions close to the isothermal ones. The polydispersion index (P_w/P_n) of the product approaches the most probable one, but depends on the difference of the initial temperature of the feed T_0 and the temperature of the boiling of the reaction mass T_{boil} (Table 4). When conditions are nearly isothermal,

Table 4. The dependent of the polydispersion index $\bar{P}_W/\bar{P}_n$ of the polymer product on the difference of the feed's initial temperature and the reagent boiling temperature $\Delta T = T_{\text{boil}} - T_0$

T_0, K	T_{boil}, K	ΔT^0	$\bar{P}_W/\bar{P}_n$
270	310	40	3.1
280	310	30	2.9
290	310	20	2.4
300	310	10	1.95
305	310	5	1.75
290	300	10	1.95
280	300	20	2.4
280	290	10	1.95

Note: $[M]_0 = 0.5$ mol/l; $[A^*]_0 = 4.5$ mol/l, $\alpha = 0$, $R = 0.08$ m.

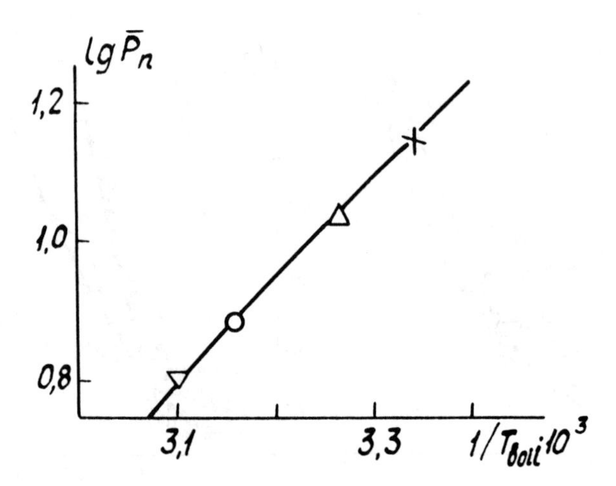

Figure 32. Dependence of lg $\bar{P}_n$ on $1/T_{boil}$ at R, $m = 0.08$; $[M]_0 = 0.5$ mol/l; $[A^*]_0 = 0.0045$ mol/l; $\alpha = 0$.

the logarithm of the polymerization degree is linearly dependent on the inverse temperature of boiling (Fig. 32).

Increasing the reaction zone radius R higher than R_{cr} when the monomer slip occurs in the vicinity of the reactor walls, the characteristic slope on the curve of the dependence of lg $\bar{P}_n$ on $1/T_0$ is levels set (Fig. 31, Curve 2), as there are zones with wide temperature ranges by the reaction zone coordinates. As a consequence, macromolecules of different sizes are formed, and the reagent boiling zone is limited by the "torch" epicenter. Since the "torch" zone covers less than 30% of the reaction volume, the inner heat removal is apparently inefficient at relatively large volumes due to the local boiling of the reagents.

This computation and modeling of the fast polymerization process with inner heat removal on account of the reagent boiling accords with the experimental data (Fig. 31).

3.3 THE OUTER HEAT REMOVAL EFFICIENCY

The peculiarities of the monomer polymerization in flows of different volume suggest that conventional methods of removing the reaction heat—in particular, external thermostating—should influence the polymerization macrokinetics and, consequently, molecular-mass characteristics of the polymer products formed in various ways. In order to provide relatively stable operation of the reactor under the conditions close to the isothermal ones, reactor-polymerizers in real production use are equipped with a complex system of external and internal thermostating, with the total surface area up to 130 m^2 [22].

Analysis of temperature variations in reaction flows in fast polymerization shows [41, 42] that the increase of the reaction zone radius R inner heat removal is missing leads to essential changes in the temperature regime of the reaction. For instance, at small radii the temperatures of the central ($R = O$) (T_0) and peripheral (T_R) flows of the reaction mixture differ little, not more than $\Delta T' = 2 \div 5°$ and only at the initial stage of polymerization (Fig. 33). With the increase of R (up to $R = 0.5$ m) $\Delta T' = 50 \div 60°$ reaches the greatest values despite the low (up to 30%wt) degrees of the monomer conversion. In this case the tempera-

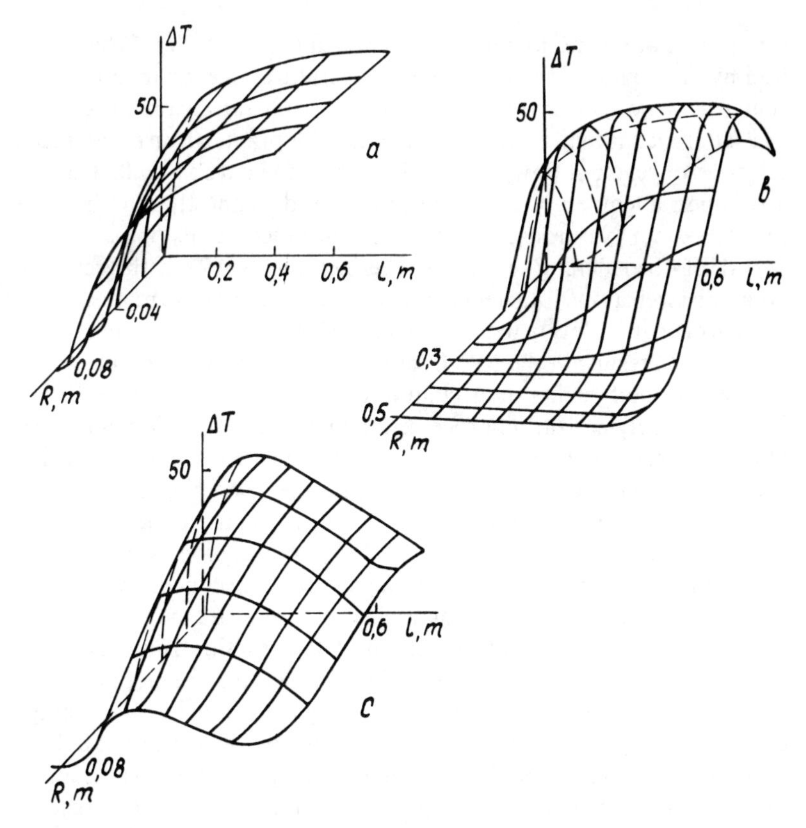

Figure 33. Surfaces of the variation of the reaction flow temperature in fast polymerization process a) $\alpha = 0$, $R = 0.08$ m; b)$\alpha = 0$, $R = 0.5$ m; c) with external heat removal: $\alpha = 50$ ($[M]_0 = 2.5$ mol/l; $[A^*]_0 = 0.0045$ mol/l, $k_p = 10^5$ l/mol·s, $k_t = 20\,s^{-1}$, $T_0 = 300$ K).

Table 5. The effect of the reaction zone radius on $\Delta T'$, $\Delta T''$ and the MWD of the product

N	R, m	$\Delta T' = T_0 - T_R (\Delta T'')$	$\bar{P}_W/\bar{P}_n$	N	R, m	$\Delta T' = T_0 - T_R (\Delta T'')$	$\bar{P}_W/\bar{P}_n$
1	0.08	0 (10)	2.0	5	0.5	9.3 (15.0)	4.2
2	0.10	7.0 (9.7)	2.5	6	0.08	1.0 (20.0)	2.2
3	0.125	9.0 (9.5)	2.9	7	0.5	19.0 (34.0)	5.9
4	0.25	11.0 (9.8)	3.4	8	0.08	1.5 (55.0)	2.5

Note: $\alpha = 0$, $k_p = 10^5$ l/mol·s; $k_t = 20\,s^{-1}$. For *N* 1–5—$[M]_0 = 0.5$ mol/l, $[A^*]_0 = 4.5 \cdot 10^{-3}$ mol/l, $T_0 = 300$ K; for *N* 6–7—$[M]_0 = 1$ mol/l, for *N* 8—$[M]_0 = 2.5$ mol/l. $\Delta T'$ corresponds to the conversion depth of 30% wt. $\Delta T'' = T_{max} - T_0$.

ture profile of the monomer polymerization is distorted and characterized by the region of the maximum temperatures occurring in the flow center with no temperature variation in peripheral regions (Fig. 33b). The latter means that the reaction does not, for practical purposes, reach the zone of reaction flow, bounded by the impenetrable wall. There is a volumetric temperature gradient (by the radius and the length of the reaction zone). Naturally, this affects the product quality, in particular, the MWD broadening (Table 5). The increase of $\Delta T'$ in the reaction volume as well as the total increase of the flow temperature related to the growth of heat evolution with the increase of the monomer content in the system (ΔT) results in the MWD broadening of the product due to the accumulation of the low-molecular fraction.

In fast polymerization processes, the polymer formation continues along the reaction zone length as well, though with the less intensity. This fact is remarkable enough—despite the very high rates of the polymerization reaction (the constant of the propagation rate of isobutylene polymerization is about $k_p = 10^6$ l/mol·s), the process proceeds at considerable distance from the entering point of the concurrent flows of the monomer and the catalyst as well. The various temperature conditions along the radius and the length of the reaction zone cause the broadening of the product MWD. The temperature reduction in the reaction zone due to heat removal through the wall at the average conversion degrees (about 50 %wt) results in decreasing the low-molecular fraction content in the final product (Fig. 34). The variation of the mean-numerical degree of polymerization $\bar{P}_n$ along the reactor length in case of outer heat removal is characterized by the minimum on the curve corresponding to the maximum heat evolved in polymerization. If, in the absence of heat removal, $\bar{P}_n$ monotonically decreases as a result of a marked growth of the low-molecular content in the polymer

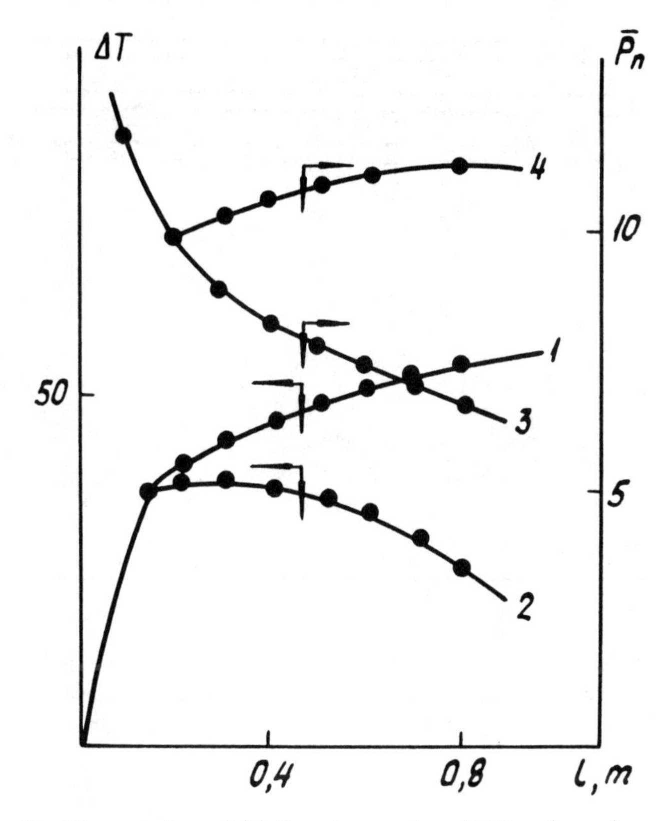

Figure 34. The variation of the flow temperature (ΔT) and number average polymerization degree ($\bar{P}_n$) along the reaction zone length (l, m) at $\alpha = 0$ (1, 3) and $\alpha = 20$ (2, 4) ($[M]_0 = 2.5$ mol/l; $[A^*]_0 = 0.0045$ mol/l, $k_p = 10^5$ l/mol·s; $k_t = 20\,s^{-1}$, $T_0 = 300$ K).

product, at already $\alpha = 10$ ($\alpha = Nu/2R$), the process is stabilized and the low-molecular fraction growth stops.

With the increase of the reaction zone volume, the external thermostating effect becomes level. At $R = 0.5$ m the heat removal stops influencing the process even with α increasing up to 100, which is connected with the monomer slip in the wall region that acts as a heat insulator.

In fast polymerization more complicated variants are also possible. In particular, the outer heat may be removed through the wall of a tubular reactor at a multi-stage catalyst supply (Fig. 35). In this case, the outside temperature is kept equal to the temperature of the cooling agent (T); the temperature in the reactor (T) is constant along the

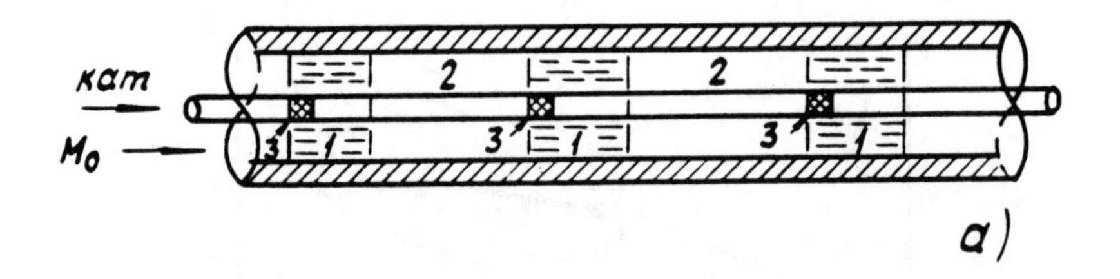

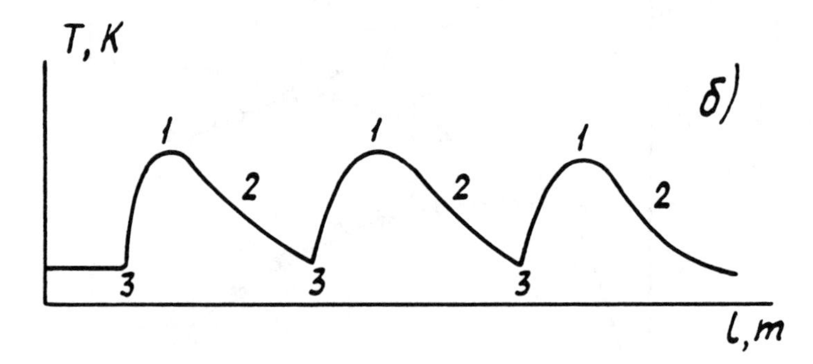

Figure 35. The temperature variation along the reactor length (l, m) with the multi-stage catalyst supply under the conditions of external heat removal: 1-reaction zone; 2-cooling zone; 3-the point of catalyst supply (diagram).

diameter; the coefficient of heat transfer through the wall (α) remains constant and the heat flow is $q = \alpha[\tilde{T} - T]$; the reaction zone is sufficiently small and all the outer heat removal is realized at the distance between the points of the catalyst supply 1_i [39, 42].

If we do not consider the longitudinal heat transfer within the reactor by turbulent heat conductivity, the temperature change (cooling down) due to outer heat removal is equal to:

$$\ln\left(1 - \frac{\Delta T_i}{T_i - \tilde{T}}\right) \cong \frac{\Delta T_i}{T_i - \tilde{T}} \cong \frac{2\alpha l_i}{\rho \cdot \bar{c}_p \cdot r \cdot V} \qquad (30)$$

where ρ, $\bar{c}_p$, V, r, l_i are the average density, average thermal capacity, the linear rate of the reaction mass flow, the reaction zone radius, and the distance between the points of the catalyst supply, respectively ($\Delta T \ll T_i - \tilde{T}$).

The change of the temperature along the reactor length at the multi-stage catalyst supply is diagrammed in Fig. 35.

Thus, the temperature in each zone will be equal to:

$$\begin{aligned} T_1 &= T_0 + q\Delta M_1 \\ T_2 &= T_0 + q(\Delta M_1 + \Delta M_2) - \Delta T_1 \\ &\ldots\ldots\ldots\ldots\ldots\ldots \\ T_i &= T_0 + q\sum_{k=1}^{i} \Delta M_k - \sum_{k=1}^{i-1} \Delta T_k \end{aligned} \tag{31}$$

Here $\Delta T_i \cong 2\alpha l_i(T_i - \tilde{T})/\rho\bar{c}_p rV$ at $\Delta T_i \ll (T_i - \tilde{T})$

(a more accurate expression is $\ln\left(1 - \frac{\Delta T_i}{T_i - \tilde{T}}\right) = -\frac{2\alpha l_i}{rV\rho\bar{c}_p}\Big)$

when there is no boiling.

At the boiling and heat removal $T_i' = T_0 + T_i - \sum_{k=1}^{i-1} \Delta T_k$, where T_i is the temperature in the i-th zone for the analogous reactor (with the same ΔM_i) at boiling but without heat removal. The value of T_i is determined in Fig. 30.

One may quantitatively evaluate the real coefficient of heat transfer α through the metal wall. In the case of outer heat removal in the pipe at the flow rate $V = 1$–10 m/s, α varies within $8400\ (2000) < \alpha < 42000\ (10000)$ kJ/m^{2}h·deg. Correspondingly, substituting the value of α in eq. (31), we will at the outer heat removal by liquid ethylene ($T = 183$ K) obtain the process temperature of 280 K for cooling down the system by 10 K, $l_i/r = 20$; at cooling by liquid ammonia ($T = 243$ K, $T_i = 300$ K), $l_i/r = 40$.

It follows from the correlation (3) that heat removal efficiency is in inverse proportion to the reaction radius at the constant linear rate or in direct proportion—at constant volumetric flow rate—of the reaction mass ($\Delta T_i = 2\alpha l_i(T_i - \tilde{T})r\pi/\rho\bar{c}_p w$, where $w = \pi r^2 V$). Consequently, to keep reactor productivity approximately constant and trying to improve outer heat removal, one should increase the reactor radius. However, that may cause the gradients of the concentrations of the reagents and temperatures at $R > R_{cr}$ to rise or the flow conditions to change from turbulent to the laminar, which would lead to deterioration of heat and mass exchange. These factors restrict the reactor radius at the top.

In conclusion, there are several possible ways of controlling the temperature field in reactors through thermodynamic factors. The boiling temperature and boiling heat depend on the nature of the solvent used. The boiling temperature depends on the pressure in the reactor. The nature of the solvent and the pressure in the reactor can cover a rather wide temperature range that may serve as an efficient means of control.

4 PROBLEMS OF THE REAGENT MIXING IN TURBULENT FLOWS

4.1 TURBULENT MIXING WITHOUT CHEMICAL REACTION

The complexity of optimum mixing of reagents in a chemical reactor plays an important part in controlling fast chemical reactions proceeding in the diffusion region. Uniform reaction mass can be obtained by mixing liquids in a turbulent flow. This is the basic operating principle of tubular turbulent reactions. The creation of optimum hydrodynamic conditions is critical, for it makes it possible to influence the character and the rate of chemical processes and, ultimately, the properties of the polymer produced. Under conditions of the diffusion-kinetic regime, the chemical reaction run is determined not only by the rate of the chemical reaction itself and its heat effect, but also by the rate of heat and mass transfer which in its turn may considerably depend on hydrodynamic conditions such as the consumption of ingredients, the rate of flow (V) and the conditions of the flow entering the reactor, the reactor geometry and size, etc. The experimental and computational data [43–45, 78] indicate that considerable improvement of the mixing procedure and increased efficiency of mass and heat diffusion may be achieved only when recirculation zones are present in the reaction volume. As a rule, in applied technology of chemical processes, these zones are created by different mechanical devices [46] and controlled flow rates and their densities [43]. Meanwhile, considerable improvement of the reagent mixing, other conditions being equal, may be obtained by changing only the geometry of a tubular turbutent reactor and the mode of the reagent supply [37].

Using the well-known equations of Navie–Stoks together with the "q–ε" equations of the turbulent model [36, 78], we considered the process of mixing two liquid-phase, chemically inert flows, containing two components (P + M and P) in five different reactors R1–R5, where only the geometry of the flow supply zone and the mode of supply differed (Fig. 29). In particular, reactors R4 and R5 have conic broadening in the initial part of the apparatus, the presence of which under the definite flow conditions may lead to the formation of recirculation zones. The value of τ_{mix} (z)—the characteristic time of mixing—has been chosen as the characteristic of mixing.

$$\tau_{\text{mix}}(z) = R^2(z) D_{T,av}^{-1}(z) \tag{32}$$

Here $R(z)$ is the reactor radius in the section z, and $D_{T,av}$ is the average by the section z value of the coefficient of effective turbulent diffusion. Figure 36 represents the distribution of the D_T values by volume for the reactor geometries studied. It follows that D_T increases (τ_{mix} decreases) for the reactors with the radial ways of flow supply ($R2$, $R3$) and especially for the reactors with conic broadening ($R4$, $R5$). For $R5$ τ_{mix} decreases by about an order, as compared with $R1$ under the given flow parameters. The increase of D_T causes quicker leveling of the concentration profile (Fig. 37). Practically the complete mixture is reached in reactors $R4$ and $R5$ on distance of one calibre.

The selection of the pulsation characteristics at the inlet section and their influence on the mixing characteristics are of considerable importance. The experimental data on this problem are usually missing. It is known that without the preturbulization of the flow at the inlet section, the value of $q_0^{1/2}/V_0$ (q_0 is the turbulent energy, V_0 is the flow speed at the inlet section) may amount to several tens of per cent. Therefore, the value of $q_0^{1/2}/V_0$ has been varied from 1 to 100%. The change of the level of initial turbulization greatly affects mixing characteristics. This influence is readily characterized by means of the miscibility coefficient γ_m:

$$\gamma_m = 1 - \frac{1}{2S}\int_s |C_m - \bar{C}_m| dS \tag{33}$$

that numerically estimates the nonuniformity of the concentration profile C_m in a certain section with the cross-section area S (Table 6).

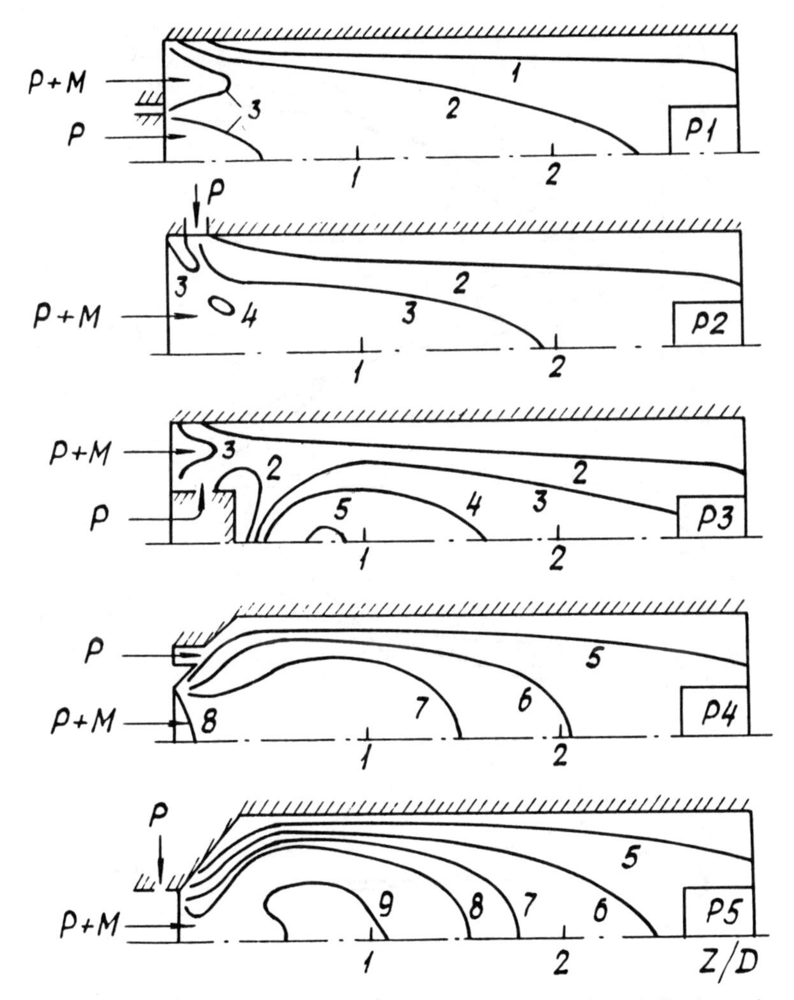

Figure 36. The diagrams of flow liquid streams and distribution of effective turbulent diffussion coefficient through the volume of five types of reactors-mixers ($[M]_0 = 0.133$ mol/l; $V_0 = 5$ m/s; $R = 0.25$ m; $T_0 = 300$ K; $q_0^{1/2}/V_0 = 50\%$) at D_T, m^2/s: 1 – 0.01; 2 – 0.02; 3 – 0.03; 4 – 0.04; 5 – 0.05; 6 – 0.10; 7 – 0.15; 8 – 0.20; 9 – 0.30.

The effect of initial turbulization on the recirculation zone, the size and intensity of which decrease with the increase of initial turbulization, is essential for reactors $R4$ and $R5$. This leads to the same reduction of D_T in the region of the recirculation zone (Fig. 38).

Choosing the best hydrodynamic conditions for the technology of chemical processes, one can essentially improve the characteristics of the

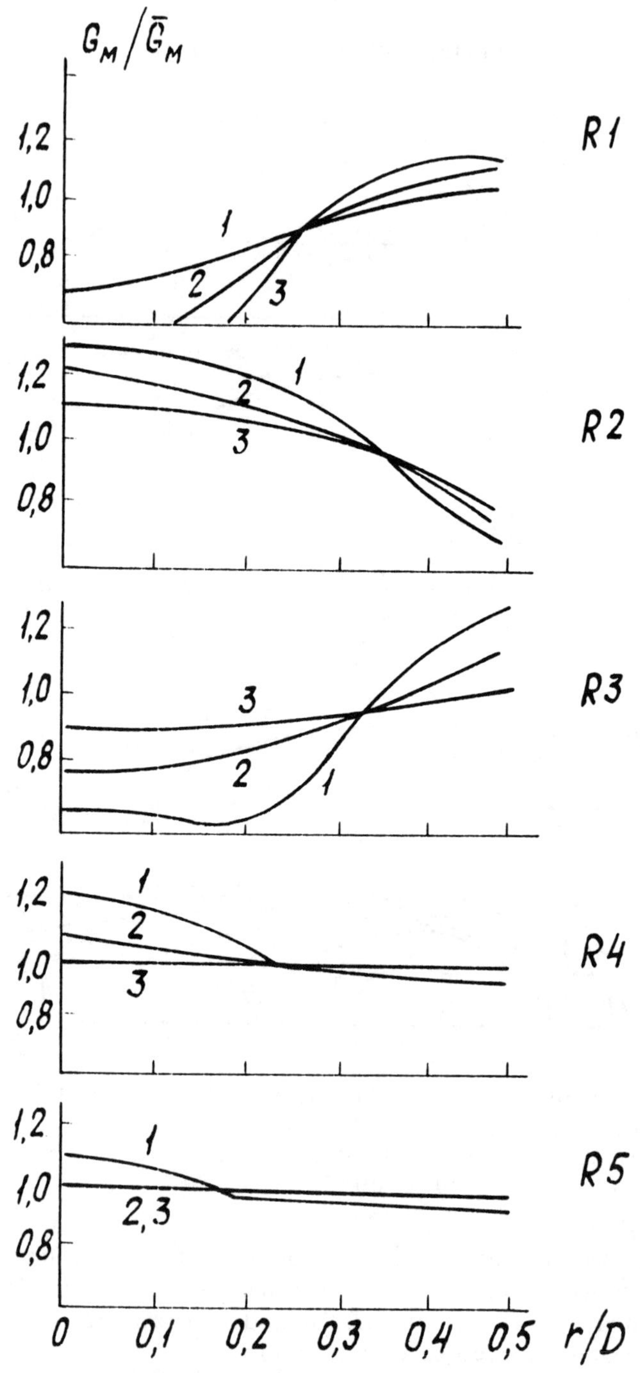

Figure 37. Radial profiles of the monomer concentration ($[M]$): z/d: 1 – 0.5; 2 – 1.0; 3 – 2.0; ($[M]_0 = 0.133$ mol/l; $V_0 = 5$ m/s; $R = 0.25$ m, $T = 300$ K; $q_0^{1/2}/V_0 = 50\%$) for reactors $R1$–$R5$.

Table 6. The dependence of the miscibility coefficient γ_{mix} on the preliminary turbulization of the flow

$q_0^{1/2}/V_0$, %	z/D	The miscibility coefficient γ_{mix}				
		R 1	R 2	R 3	R 4	R 5
	1	0.917	0.938	0.931	0.984	0.993
	2	0.944	0.963	0.959	0.995	0.998
100	4	0.969	0.982	0.978	0.999	1.000
	6	0.979	0.989	0.985	1.000	1.000
	1	0.907	0.930	0.924	0.985	0.993
	2	0.944	0.957	0.955	0.996	0.998
75	4	0.960	0.978	0.976	0.999	1.000
	6	0.972	0.986	0.984	1.000	1.000
	1	0.895	0.920	0.912	0.984	0.993
	2	0.920	0.949	0.948	0.996	0.999
50	4	0.946	0.972	0.976	0.999	1.000
	6	0.960	0.982	0.984	1.000	1.000
	1	0.878	0.905	0.902	0.983	0.993
	2	0.897	0.937	0.945	0.995	0.999
25	4	0.921	0.964	0.974	0.999	1.000
	6	0.936	0.976	0.984	1.000	1.000
	1	0.865	0.981	0.889	0.980	0.991
	2	0.875	0.928	0.942	0.995	0.998
10	4	0.892	0.957	0.974	0.999	1.000
	6	0.903	0.970	0.985	1.000	1.000
	1	0.853	0.881	0.876	0.978	0.990
	2	0.855	0.931	0.934	0.993	0.996
1	4	0.858	0.964	0.972	0.999	1.000
	6	0.860	0.977	0.984	1.000	1.000

transfer processes, making it possible to control the quality of the product obtained in fast polymerization processes.

At the same time, it turned out that mixing the flows of liquids that differ in density and especially in viscosity is a problem. Video-filming has revealed that "water-glycerine" is not the only difficult mix. It is limited by impenetrable wall with R-15 mm at the linear flow speed (V) rate of 0.5–0.8 m/s and the concurrent reagent supply. But "water-concentrated sulfuric acid" does not mix well either.

These experimental facts relate to problems of mixing liquids with various viscosities and densities in the flow.

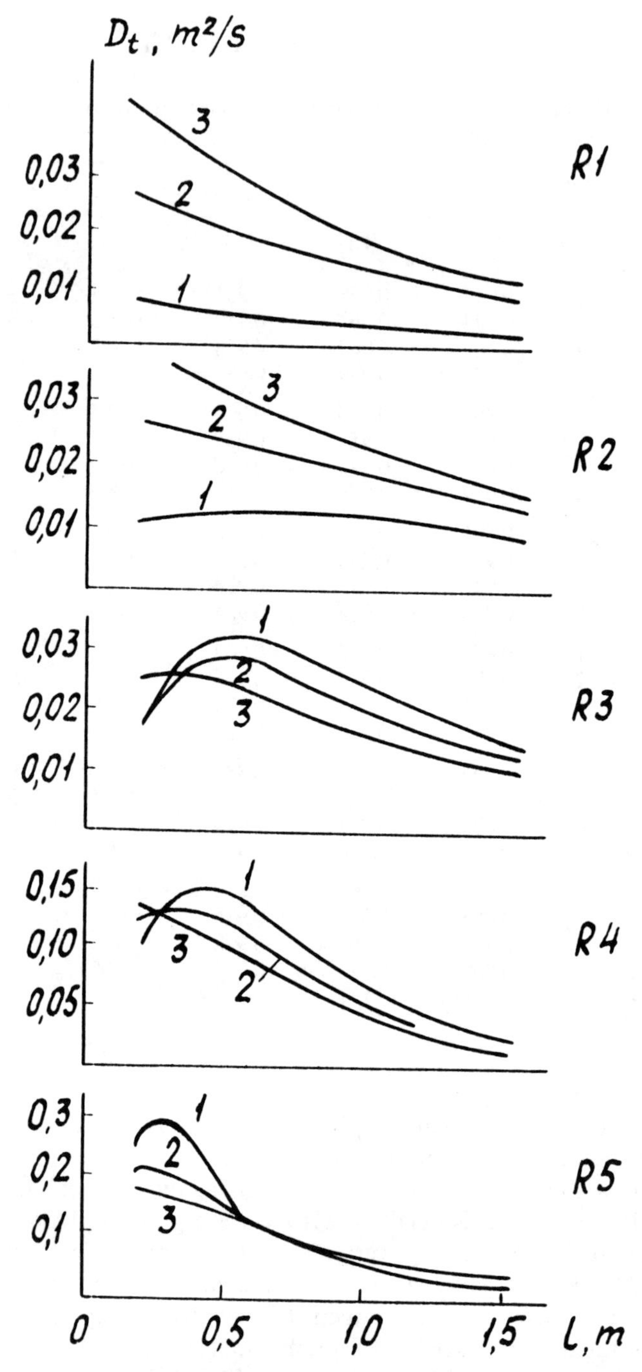

Figure 38. The dependence of the mean in the section coefficient of turbulent diffusion D_T on the way of mixing anf flow preturbulization $q_0^{1/2}/V_0$ %: 1–10; 2–50; 3–100; ($[M]_0 = 0.133$ mol/l; $V_0 = 5$ m/s; $R = 0.25$ m; $T_0 = 300$ K.) for reactors $R1$–$R5$.

4.2. THE INFLUENCE OF TURBULENCE ON THE EFFICIENCY OF MIXING THE LIQUID FLOWS WITH VARIOUS DENSITIES AND VISCOSITIES

Most fast chemical reactions proceed on the border layer of the reaction flow when two reagents do not mix well. It is easy to simulate mixing two liquids by the example of obtaining emulsions in the interaction of two non-mixing liquid flows, including liquids of differing densities and viscosities. This simulation is of great importance.

In practice, this simulation is important for the processes of electrophilic polymerization of olefins at the catalyst supply stage (by dissolving $AlCl_3$ in the chlororganic compounds etc.), in diluting concentrated sulfuric acid with water, in washing out the catalyst with the alkali solutions in water, in stoppering the process of polymerization, etc.

For heterogeneous systems in which water is the dispersing medium and the dispersed phase appeared as the organic liquid colored by Sudan-III (hexane, ρ is 0.66 g/cm^3, η is 0.3 cp; octane, $\rho = 0.73$ g/cm^3, $\eta = 0.73$ cp; the mixture of isooctane with CCl_4, $\rho = 1.04$ g/cm^3; dichloroethane, $\rho = 1.26$ g/cm^3, $\eta = 0.83$ cp; chloroform, $\rho = 1.43$ g/cm^3, $\eta = 0.56$ cp, CCl_4; $\rho = 1.60$ g/cm^3, $\eta = 0.93$ cp), the mixing efficiency is determined by the surface phase area. That area can be readily evaluated by the distribution of drops by their size (d_d) and the percentage (n) of drops with $d_d < 1$ mm. The effect of turbulence has been estimated by the coefficient of turbulent diffusion (D_T), as defined for each particular mixing apparatus.

The coefficient of turbulent diffusion has been calculated on the "q–ε" turbulence model [38, 78]. The standard form of the equation is as follows:

$$a_\phi\left[\frac{\partial}{\partial z}\left(\phi\frac{\partial\psi}{\partial r}\right)-\frac{\partial}{\partial r}\left(\phi\frac{\partial\psi}{\partial z}\right)\right]-\frac{\partial}{\delta z}\left[b_\varphi\frac{\partial(c_\phi\phi)}{\partial z}\right]-\frac{\partial}{\partial r}\left[b_\phi\frac{\partial(c_\phi\phi)}{\partial r}\right]+d_\phi=0 \tag{34}$$

The coefficients a_ϕ, b_ϕ, c_ϕ, d_ϕ and the notation of values contained in the equation are given in [39, 47].

In addition, the mixing efficiency to a considerable extent depends on the relative variation of the coefficient of turbulent diffusion along the reactor axis (D_{rel}) that is determined according to the formula:

$$D_{\text{rel}}=\frac{|D_0-D_k|}{l_k} \tag{35}$$

D_0 and D_k are the coefficients of turbulent diffusion at the reactor inlet and outlet respectively; l_k is the distance along the reactor axis in calibers.

Under experimental conditions, emulsions with concentrations from 1 to 10% have been produced and the rate of the dispersed phase supply has exerted no marked influence on the dispersion index of the system.

In case the constant D_T ($D_{rel} > 0.01$) is not provided in the flow direction, the dispersing of the liquid is characterized by extremely low effectiveness.

At the same time, tubular turbulent apparatus proved to be highly efficient in dispersing non-mixing flows with the constant value of the turbulent diffusion coefficient ($D_T = 0.045 - 0.120\,m^2/s$). In this case, the rate of the supply of dispersing medium plays a decisive part in obtaining higher degrees of dispersion. In particular, at the rates of flow lower than 0.6 m/s (when the radius of turbular apparatus R is 10 mm), transitional flow conditions occur that are characterized by the presence of rather large emulsion drops ($d_d = 1.5 - 2$ mm). Increasing the flow rates up to 0.8 m/s, turbulent flow conditions, accompanied by a sharp increase of the index of the system dispersion, have been observed (Fig. 39). Further

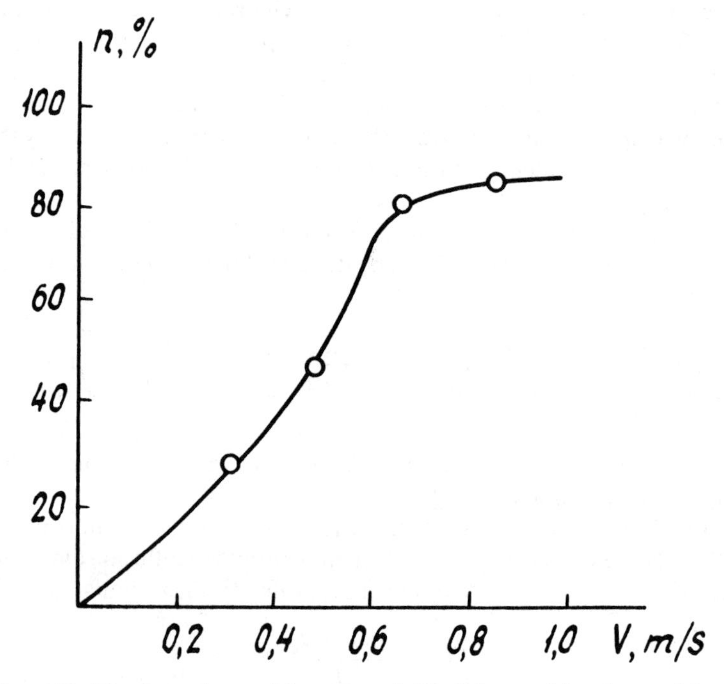

Figure 39. The dependence of the content (n%) of the emulsion drops of the size $d < 0.8$ mm on the linear rate (V, m/s) of the dispersion medium supply (water) in dispersing hexane in a tubular turbulent reactor with $D_T = 0.065\,m^2/s$.

increases in the flow speed (up to 1 m/s) result in the transition of the process to the region of the developed turbulent flow and is characterized by a higher degree of dispersion and by uniformity of the system (98–100% of the drops are of the size $d = 0.55 \pm 0.15$ mm).

At the fixed flow speed, the efficiency of the system dispersion can be controlled by varying the value of D_T, with further stabilization by means of intense turbulent vortex.

The turbulent apparatus with D_T of $0.065 - 0.080\,m^2/s$ (Fig. 40) proved most efficient. Both the reduction and increase of the values of D_T as compared to those mentioned above, lead to decreases in the dispersion degree of the system.

It was expected that using tubular turbulent reactors characterized by the variable value of D_T along the reactor axis (D_T increases or decreases with a certain amplitude or step-by-step at $D_{rel} < 0.01$) should have resulted in the growth of the efficiency of the emulsification processes. However, in this case the degree of dispersion does not markedly change in comparison with the apparatus with $d_T = \text{const} = 0.080\,m^2/s$ (Table 7).

The value of the coefficient of turbulent diffusion D_T as well as the character of the variation along the reactor axis also influences the mixing efficiency.

In particular, despite the unlimited solubility of the concentrated sulfuric acid ($\rho = 1.64\,g/cm^3$, $\eta = 18.2$ cp) and glycerine $\rho = 1.26\,g/cm^3$, $\eta = 1490$ cp) in water, applying the tubular turbulent apparatus characterized by the variable coefficient of turbulent diffusion D_T along the reaction zone length, results in the retention of the surface of the boundary of two phases. The sulfuric acid (glycerine) first forms pools at the botom of a tubular apparatus and then is distributed non-uniformly in the water flow in the form of "plaits," clots etc.

One increases the mixing efficiency of flows not so much by increasing the linear flow speed (V) as by stabilizing D_T at the difinite level in the flow direction.

It should be noted that the correlation of the component viscosities exerts especially great influence on the formation of thinly dispersed emulsions in dispersing two non-mixing liquids. The fragmentation of the viscous component with the formation of emulsion even at the relatively small difference in viscosities of the dispersing medium and the dispersed phase ($\eta_{disp.m}:\eta_{disp.ph} = 1{:}3$) is very problematic. This, in particular, is distinctly shown in the example of dispersing the flow of the solution of stereoregular 1–4 cis polyisoprene rubber (ISR-3) in heptane of various concentration colored by Sudan-III. The rate of dispersed phase flow V is 0.01 m/s. The rate of the dispersing medium (water) V is

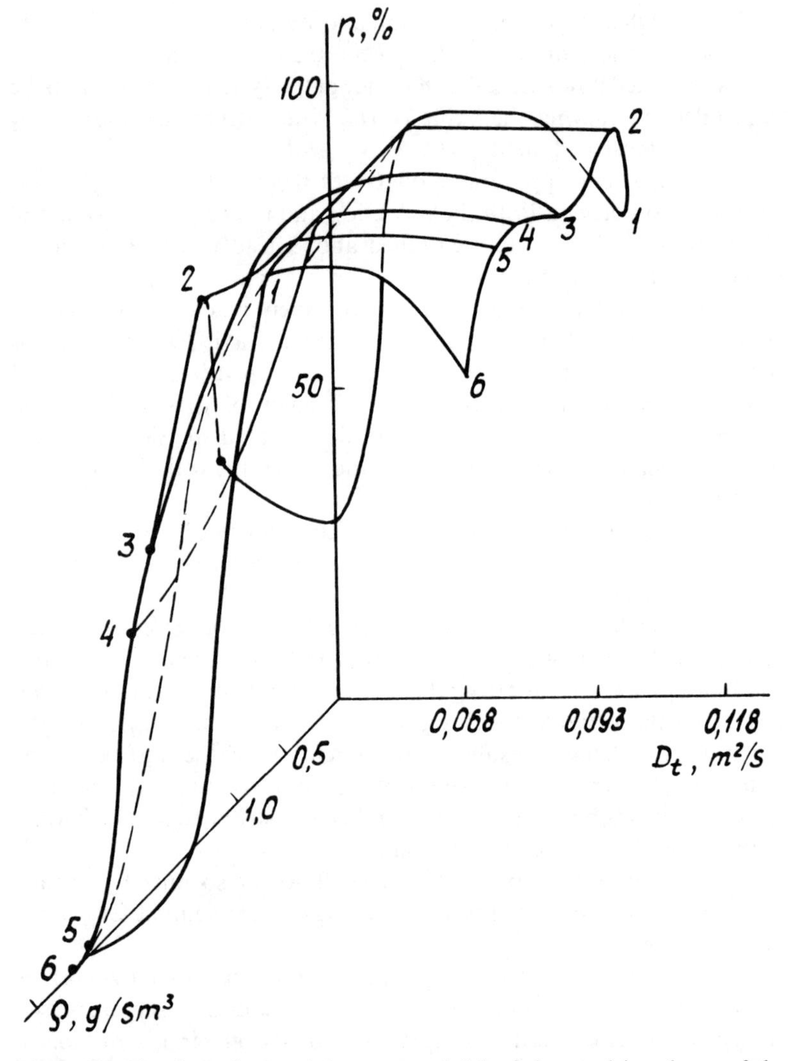

Figure 40. The dependence of the content (n%) of the emulsion drops of the sizes d < 0.8 mm on the density of the dispersion phase supplied (ρ, gm/cm^3) and the flow turbulence coefficient (D_T, m^2/s): 1 – hexane; 2 – octane; 3 – mixture of isoctane with carbon tetrachloride; 4 – dichloroethane; 5 – chloroform; 6 – carbon tetrachloride.

Table 7. The influence of variation of D_T in a tubular apparatus—on the dispersion effectiveness of non-mixing flows

D_T, m²/s	D_0, m²/s	D_k, m²/s	D_{rel}	n, %
$D_T = \text{const}$	0.065	0.065	0.000	90
	0.080	0.080	0.000	93
	0.100	0.100	0.000	90
	0.120	0.120	0.000	85
$D_T \neq \text{const}$	0.080	0.120	0.006	95
	0.120	0.080	0.006	95
	0.120	0.065	0.008	90
	0.120	0.055	0.010	60
	0.045	0.005	0.500	55

Table 8. The density (ρ), dynamic viscosity (η) and the surface tension (σ) of the solutions of stereoregular polyisoprene rubber (ISR-3) in heptane (295 K)

N°	C, (%wt)	$\rho \pm 1$, (kg/m³)	η, (cp)	$\lg \eta$	$\sigma \cdot 10^3$ (N/m)
1	11.7	720	1.27	0.104	24
2	14.2	720	2.79	0.446	25
3	16.0	719	4.98	0.697	24
4	17.7	719	8.47	0.928	25
5	19.6	717	15.18	1.181	25
6	20.0	718	17.59	1.245	25
7	21.6	718	28.00	1.447	25
8	23.6	—	52.65	1.721	25
9	24.8	723	77.88	1.891	25
10	27.9	732	205.49	2.313	25
11	28.6	—	256.69	2.409	25
12*	0.0	688	0.29	1.06	25

12* are the characteristics of pure heptane.

0.7–1.1 m/s. The rubber solutions in heptane, independent of their concentrations, are characterized by the constant surface tension and density (Table 8).

Since the second component was introduced in relatively small quantities (less than 1%), its contribution to the change of the total linear speed of the flow may be neglected. The emulsification efficiency was estimated by the content (n%) of the fraction with drops of sizes $d \leqslant 0.8$ mm and by their distribution by size. The minimum size of the particles, which was recorded experimentally by photography, amounts to 0.25 ± 0.05 mm.

The dispersion of the viscous component in the system "water-solution of stereoregular polyisoprene rubber" (ISR-3) in heptane was realized in tubular turbulent reactors.

The dispersion of the viscous component in the system "water-solution of stereoregular polyisoprene rubber" (ISR-3) in heptane was realized in tubular turbulent reactors of the type $R5$ (Fig. 36), differing in the quantity of the diffuser angle opening (α) in the range from 0° to 60°.

Figure 41(a) represents the typical histogram of drop distribution by size in dispersing in tubular reactor $R5$ with the angle $\alpha = 60°$ at different rates given for the dispersing medium.

The increase of the rate of the flow supply causes the distribution narrowing by drop size and transition from the polymodall to mono-modall distribution. The growth of the linear rate of the dispersed phase supply at the apparatus outlet, exceeding 1 m/s, considerably increases the content of thinly dispersed fraction in the emulsion ($d \leqslant 0.8$ mm), up to 96–100% (Fig. 41b). The computation of the Reynolds number at the inlet to a tubular apparatus, using the approximation in which the second component exerts no marked influence on the supply rates, has shown that up to the flow rate of 0.9 m/s transitional flow conditions exist, and at $V > 1.0$ m/s they are followed by the turbulent ones (Re = 11000, Fig. 41b).

The character of the viscosity effect of the component being dispersed essentially depends not only on the liquid rate in the flow but, to a greater extent, on the geometric parameters of the mixing zone. For a tubular turbulent apparatus with the angles $\alpha = 0°$ and $\alpha = 15°$, this effect is complex throughout the whole investigated interval of variation of the dispersed phase viscosity, the scatter of the points being great and having an error rate up to 20%. The effectiveness of the formation of the uniform thinly dispersed emulsion is not high. The content of the drops with $d \leqslant 0.8$ mm does not exceed 50%, even under the turbulent conditions of the main flow ($V = 1.1$ m/s, Re = 11000); that is, it is evidently related to the insufficient turbulization of the flow by the viscous dispersed phases for the efficient fragmentation in such types of apparatus. Increasing the angle α up to $30° \div 60°$ markedly increases the efficiency of the system dispersion (Fig. 42). In comparison with the fact of using the mixers with $\alpha = 0$–15°, the content of the fraction with drop size of $d \leqslant 0.8$ mm increases up to 80% at the same rates of supply in applying the apparatus with $\alpha = 30°$–60°. Simultaneously, considerable narrowing of the drop distribution by size is observed, that being of great importance.

Under transitional conditions of the main flow, the dependence of the drop fragmentation on the dispersed phase viscosity is extreme

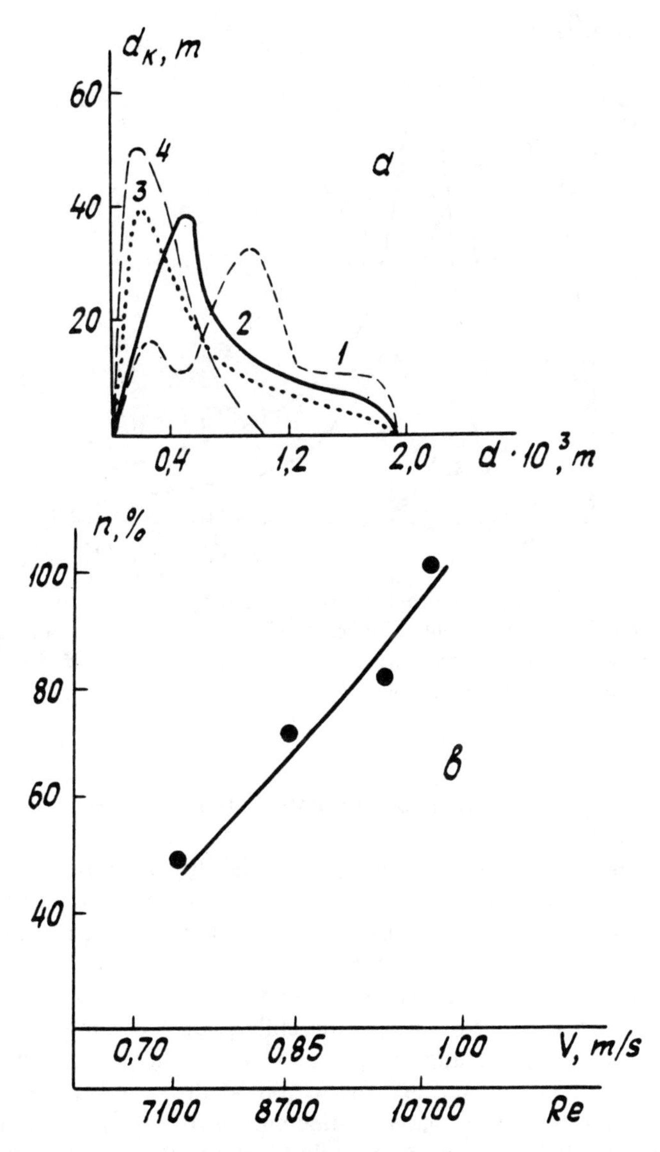

Figure 41. Distribution of drops by sizes (a) and content of fine dispersed phase with the drop sizes of $d < 0.8 \cdot 10^{-3}$ m (b) on rate flow in the reacor of $R5$ type of dispersium medium (water) V, m/s: 1 – 0.7; 2 – 0.9; 3 – 1.0; 4 – 1.1. The dispersion phase is the polyisoprene synthetic rubber (ISR-3) solution in heptane: concen tration 24.8 %wt, $\rho = 723$ kgm/m^2, $\eta = 77.9$ cp, $\sigma = 0.025$ N/m.

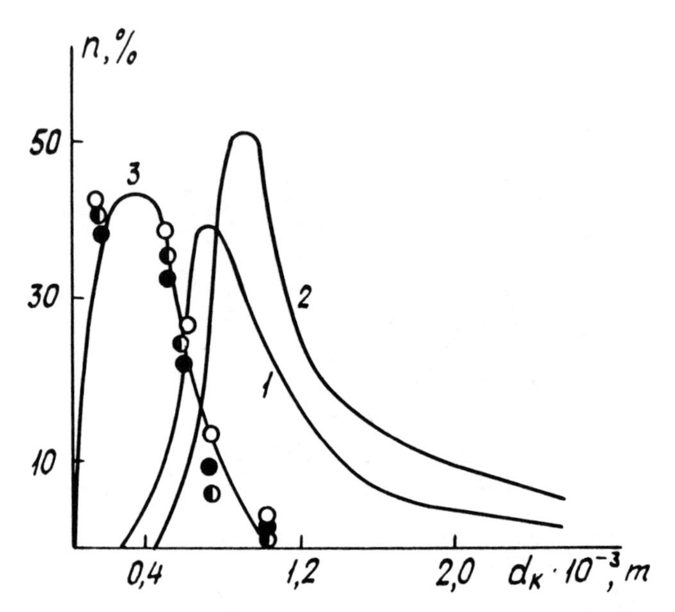

Figure 42. Distribution of drops by size by dispersing of the polyisoprene synthetic rubber (ISR-3) solution in heptane in a rube tubular reactor with the angle of the diffusor opening: α^0: 1 – 0; 2 – 15; 3 – 30 ÷ 60 (0–30; ●–45; ◐–60). Characteristic of dispersion phase give in Fig. 43.

(Fig. 43a), with the experimental points scatter up to 20% and more. At secured turbulent character leak of the main stream (water) (Fig. 43b), the efficiency of the viscous liquid dispersion (the emulsion quality) markedly increases. Under these conditions (at the dispersed phase viscosity $\leqslant 200$ cp) high-quality emulsions are always formed, with the thinly dispersed fraction content ($d \leqslant 0.8$ mm) of $85-98\%$, ($d_{av} = 0.4 \pm 0.05$ mm) and narrow drop distribution by size.

For the dispersed phase with still greater dynamic viscosity ($\eta > 250$ cp), we see a change in the character of the liquid emulsification. The viscous component (the dispersed phase) does not break down to drops, but is broken to "plaits" of different size: $(0.6–1.0) \cdot 10^{-3}$ m thickness and $(2–3) \cdot 10^{-3}$ m long for a tubular tubulent apparatus with the diffuser angle opening $\alpha = 30°–60°$, and of $(2–4) \cdot 10^{-3}$ m thickness and $(10–15) \cdot 10^{-2}$ m long for those with $\alpha = 0°–15°$.

As under experimental conditions, the surface tension and the density of rubber solutions in heptane were not significantly changed, the critical conditions for maintaining the sperical shape of the drops are

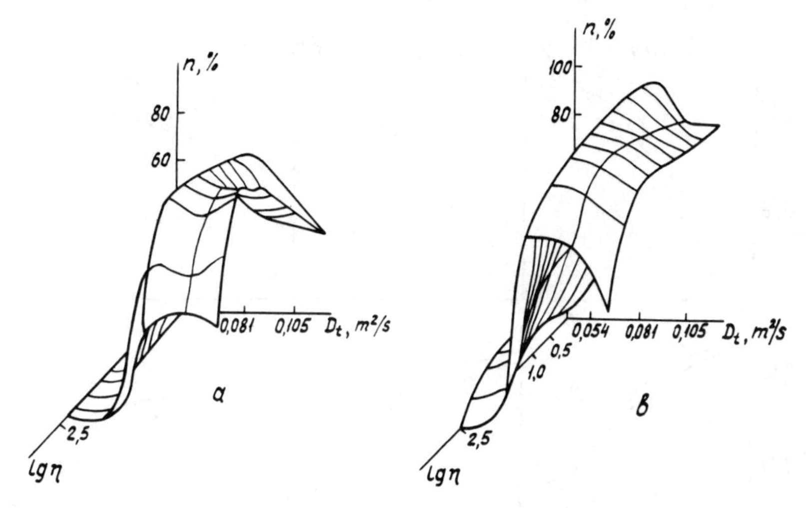

Figure 43. The dependence of the content of fine-dispersed phase (n%) with the drop sizes of $d \leqslant 0.8 \cdot 10^{-3}$ m on the viscosity of the emulsifying component (8–24% solution of ISR-3 in heptane) and rates of the dispersion medium supply, V, m/s: $a = 0.7$, $b = 1.1$.

determined only by the Reynolds number for the dispersed phase: $Re_{ph} = \rho_{ph} \cdot V \cdot R / \eta_{ph}$ and the role of the Weber and Zote criteria is not significant [48].

The calculation of R_{ph} for the drop size of $d \sim 0.3$ mm at the linear flow speed $V \sim 1.1$ m/s gives the critical value for the viscosity of the component being dispersed (at $Re_{ph} \cong 1$) of 240 cp. For the solutions with lower viscosity ($Re_{ph} > 1$), their fragmentation to spherical drops occurs, and for solutions with the higher one ($Re_{ph} < 1$), the drops are detached in the form of clots of the viscous component ("plaits"), but the latter are not broken down.

4.3 THE INFLUENCE OF TURBULENT MIXING OF FLOWS ON HOW FAST POLYMERIZATION PROCEEDS

The character of mixing notably influences the proceeding of fast polymerization reactions [39]. The dependence of the miscibility coefficient γ_m, numerically characterizing the uniformity of the monomer concentration profile, on the reaction zone length for various types of mixing and on the depth of the monomer conversion is given in Fig. 44.

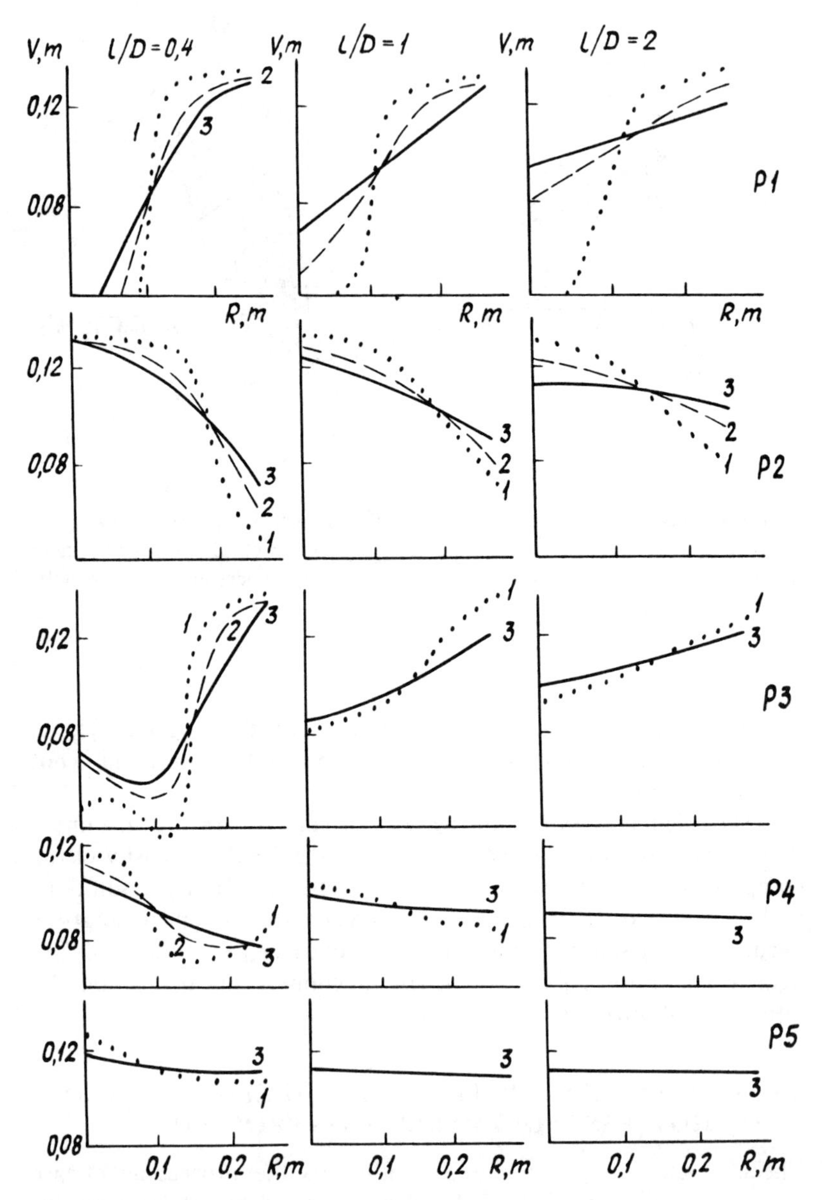

Figure 44. The distribution of the monomer mass concentration depending on the ways of mixing and preturbulization level: $V_{M_0} = V_{A_0^*} = 5$ m/s; l = 0.1 m, $R = 0.25$ m; 300 K; 1 – 10%; 2 – 50% in fine type of reactors-mixer 3 – 100%.

At the intense proceeding of a chemical reaction, the active particle transfer in a number of cases may turn out to be much lower than the rate of the chemical reaction itself ($\tau_{ch} \gg \tau_{mix}$). That can cause considerable change in the monomer concentration due to its "burning out" in the active zone.

This naturally leads to reduction of the monomer conversion depth at the initial stage and to the growth of the concentration gradient (corresponding to the minimum one on the curves of γ_m variation). All these phenomena cannot be compensated by the monomer influx from outside. The similar effects, obviously, are explained by the periodic temperature oscillations in the isobutylene polymerization process (Fig. 1) [23]. The intensification of turbulence influences the temperature field of the reaction. So, at the maximum transfer of the reaction heat through the wall, thermostatic effect manifests itself to a considerably greater extent. This results in the significant decrease of temperature in the reaction zone with the simultaneous increase of the monomer conversion (1.5–2 times). The growth of the average MW and narrowing of the MWD follows the temperature reduction.

In [36], the influence of mixing efficiency in fast polymerization process in the liquid phase under the conditions of non-isothermal flow movement in the axisymmetric reactor by two models is compared. The former model admits the constant coefficient of turbulent diffusion D_T along the reaction volume (Model I). The later model—described by the famous stationary equation of Navie–Stoks together with the "q–ε" equations of the turbulence model (Model II) [39]—involves two ways of the reagent supply: concurrent and radial. Within these models the change of the method of the catalyst supply is simulated by the change of the coefficient of the turbulent diffusion D_T.

Typically, in polymerization, including that of isobutylene, great change is possible for medium rheology and the turbulent flow characteristics. However, in the given case, low degrees of polymerization and sufficient dilution were considered. The initial monomer concentration was 2.5 mol/l. In this case the relative viscosity of the polymer solution is 2–3 [49], while the effective turbulent viscosity is 3 orders more than that of the solvent. Under these conditions the polymerization effects on the medium rheology and turbulent characteristics are insignificant and may be neglected.

The applied models, the computation algorithm and the boundary conditions are given in detail in [39]. Model II is based on the solution of the stationary equations of continuity, energy and concentration for the averaged values:

$$\frac{\partial}{\partial x_j}(\rho U_j) = 0 \tag{36}$$

$$\frac{\partial}{\partial x_j}(\rho U_j U_i) - \frac{\partial}{\partial x_j}(\tau_{ij}) + \frac{\partial \rho}{\partial x_i} = 0 \tag{37}$$

$$\frac{\partial}{\partial x_j}(\rho U_j \phi) + \frac{\partial}{\partial x_j}(J_{\phi j}) - R_\phi = 0 \tag{38}$$

where τ_{ij} is the viscous stress tensor, $\phi = (m_s, h, q, \varepsilon)$ are the mass concentration of the s-th component, the enthalpy of stagnation, the turbulent energy, the rate of the turbulent energy dissipation respectively, J_ϕ is the flow density vector of the values ϕ, R_ϕ is the source.

In the modeling of the viscous stress tensor and the ϕ value flows, the conception of gradients has been used. In order to close the system of hydrodynamic equations (36)–(38), the "q–ε" model has been taken according to which turbulent viscosity is $\mu_t = C_D \rho q^2/\varepsilon$, where C_D is an empiric constant. The system of differential equations of the model expressed in the variables of the vorticity—the flow function has been approximated by the finely-differenced equations on the straight grid. For convectional terms, the asymmetric difference scheme, oriented against the flow, has been used.

The initial level of the turbulence pulsations is chosen to be 50%, which agrees with the experimental data [50–52]. The kinetic model of the isobutylene polymerization process ((VI)–(X)) has been used. The physical and thermophysical parameters for the reaction components have been taken from [2, 28, 53]. The result of calculations is given in Fig. 45–48 and Table 9. The analysis of the data, obtained for the concurrent and radial ways of the catalyst supply in monomer stream, shows that, all other conditions being equal, the turbulization of the flow increases and the mixing is improved in the latter case. The turbulent diffusion coefficient D_T increases (Fig. 47), the reaction zone is "compressed," and the nomomer conversion increases (Fig. 46). The molecular characteristics of polymers are also improved in the fast process of polymerization.

The comparison of the calculated results for the two models shows their sufficient agreement. Thus, within Model I, used for the process investigation, one manages to describe its main peculiarities. Therefore, for most calculations there is no urgent necessity to apply more the powerful and tedious Model II. However, for the computation of a

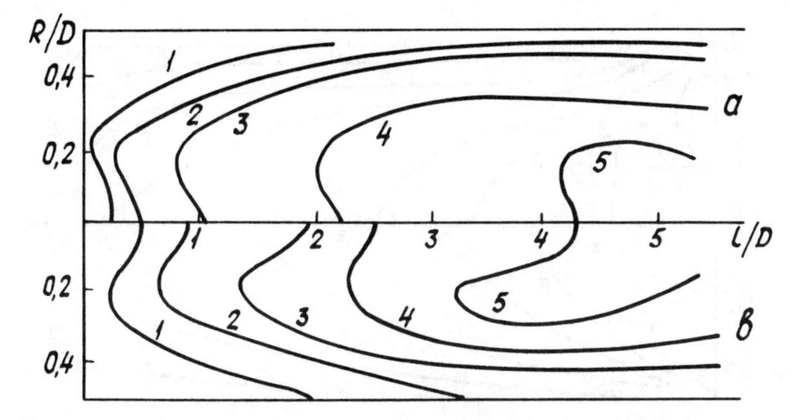

Figure 45. Temperature fields at isobutylene polymerization in the cocurrent supply of the monomer and catalyst, computed on Model I (a) and Model II (b) ($[M]_0 = 2.5$ mol/l; $[A^*]_0 = 0.01$ mol/l; $T_0 = 200$ K; $D_T = 3 \cdot 10^{-4}$ m^2/s (for Model I); T, K: 1 – 210; 2 – 230; 3 – 250; 4 – 270; 5 – 276.

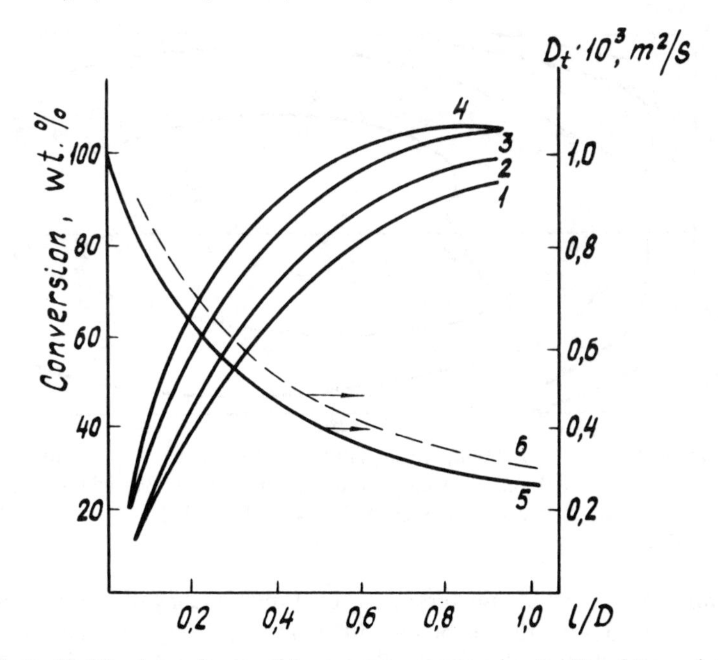

Figure 46. The dependences of the monomer conversion (1–4) and D_T variation (5, 6) on the reaction zone length at different flow directions of the catalyst (1, 2, 5-cocurrent; 3, 4, 6-radial), obtained by Model I (2, 4) and Model II (1, 3, 5, 6). D_T (in the radial way of the catalyst supply) $= 6 \cdot 10^{-4}$ m^2/s; ($[M]_0 = 2.5$ mol/l; $[A^*]_0 = 0.01$ mol/l; $T_0 = 200$ K).

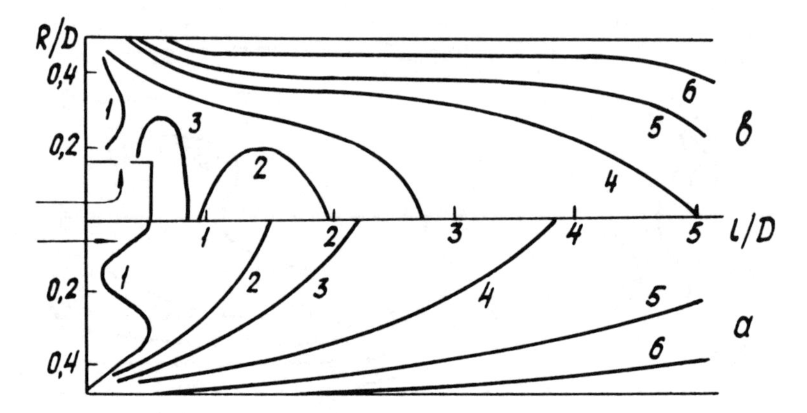

Figure 47. The distribution of the values of the turbulent diffusion coefficient (D_T) in the reacting flow volume at different ways of the reagent supply: α-concurrent, b-radial ($[M]_0 = 2.5$ mol/l; $[A^*]_0 = 0.01$ mol/l; $T_0 = 200$ K).

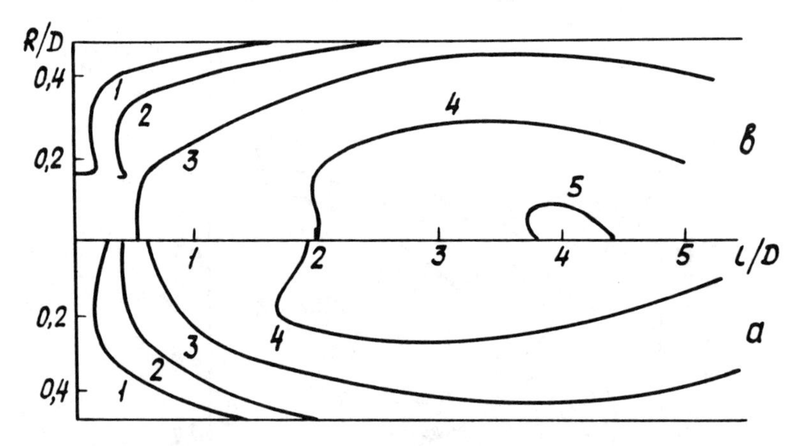

Figure 48. Distribution of temperatures in the flow of the reacting monomer in the radial method of the catalyst supply into monomer due to Model I(a), Model II(b), $D_T = 6 \cdot 10^{-4}\,\mathrm{m^2/s}$ (for Model I) ($[M]_0 = 2.5$ mol/l; $[A^*]_0 = 0.01$ mol/l; $T_0 = 200$ K).

number of concrete parameters (D_T, q, etc.) and for using of Model I with the real values of the turbulent diffusion coefficient as well as for establishing the flow dynamic's effect on the character of the chemical process, it is necessary to use Model II.

Thus, by changing the method of reagent supply, heat intensification and mass transfer, one can effectively influence the base parameters of fast polymerization processes, varying the temperature field of the

Table 9. The results of calculations by Models I and II for the concurrent and radial ways of the catalyst supply

Parameters	Model I		Model II	
	concurrent	radial	concurrent	radial
Conversion, %wt	0.99	0.99	0.97	0.99
T_{max}, K	280	279	276	274
$T_{av.max}$, K	260	260	264	262
$\bar{P}_n$	46	50	40	44
$\bar{P}_W$	234	260	246	267
$\bar{P}_z$	655	700	628	682
$\bar{P}_W/\bar{P}_n$	5.05	5.03	6.15	6.08
$\bar{P}_Z/\bar{P}_W$	2.80	2.90	2.55	2.55

Model I—the concurrent way of the catalyst supply, $D_T = 6\cdot 10^{-4}\,m^2/s$; the radial way—$D_T = 6\cdot 10^{-4}\,m^2/s$, $T_0 = 200$ K.

reaction and molecular mass characteristics of the polymer product correspondingly.

However, the intense effect on the temperature field of the reaction may be exerted by other ways, as well as varying hydrodynamic conditions. The value of the temperature variation also depends on the depth of the process proceeding, namely, on the amount of the monomer having been polymerized or on the amount of the catalyst supplied, especially when the catalyst is introduced into the reaction zone by small portions at several points (see Chapt. 6 for a discussion of the zone model of the reactor).

4.4 THE INFLUENCE OF TURBULENT MIXING ON VERY FAST POLYMERIZATION WITH THE LOW MOLECULAR COMPOUNDS INTERACTION IN THE FLOW

One of the most essential results in very fast liquid-phase polymerization process, when the reaction characteristic time is comparable or less than that of the reaction components transfer ($\tau_{ch} < \tau_{mix}$), appears to be the existence of several different macroscopic regimes of the process that vary in the structure of temperature fields and the reagent concentrations (the reaction fronts) [31, 32, 54] (Section 2.1, Chap. 2).

The first type of reaction front is the local "torch"—like one (B and C) with the gradient of temperatures and the concentrations of ingradients, the size of the "torch" being determined by the correlation of two

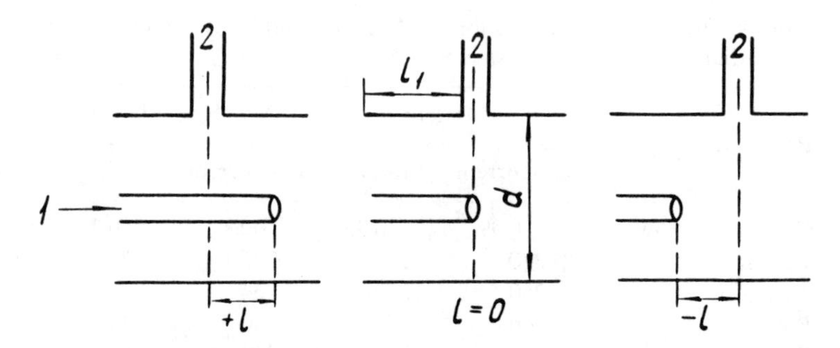

Figure 49. The diagram of the zone of the reagent supply into a tubular turbulent reactor.

competing processes—the mixing (diffusion) of the reagent flows and the active site termination (including growing polymer molecules) (Fig. 15, c–f).

The second type with no analogues represents the planar reaction front (A), where the surfaces of the equal monomer concentrations and active sites form a plane perpendicular to the reaction axis, determining the possibility of quasi-ideal displacement in turbulent flows (Fig. 15a, b).

The formation of different macroscopic front types in running fast chemical reactions is, as it turned out, a general phenomenon and extends to practically any very fast chemical processes, including those of interaction of low-molecular compounds as well. This is clearly shown on the boundary of the color appearing in the intere action of KSCN and $FeCl_3$ [55–57]. For the supply of reagents 1 and 2 at the linear flow speeds $V_1 = 0.2–7$ m/s and $V_2 = 0.1–0.4$ m/s, branch pipes of different types were used (Fig. 49).

To supply reagent 1, the connecting branch with open exit end (variant A) and the connecting branch with closed exit end but radial perforation along the tube exit of reagent 1 (variant B) were used when varying the points of supply of reagents 1 and 2 along the reactor axis. The flow expenses changed in the range of 0.4–1.4 m/s. Experimental conditions included provided turbulent flow conditions in the reaction zone at all given speeds of the supply of reagents 1 and 2 ($Re > 2 \cdot 10^3$). This is distinctly demonstrated in [38] while evaluating D_T within the framework of the "q–ε" model by Navie–Stoks [39].

Depending on the method of supplying reagents 1 and 2 and on the relationship between the linear flow speeds V_1 and V_2, three basically different structures of the reaction fronts can form.

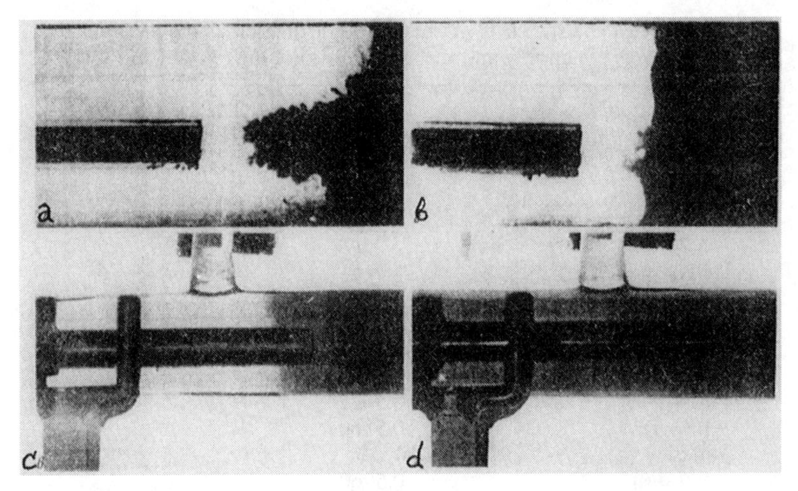

Figure 50. The reaction fronts in mixing of turbulent flows of KSCN and $FeCl_3$ in a tubular reactor; a – the method $A(+l)$, $V_1/V_2 = 1.0(\times 1.35)$; b – way $A(+l)$; $V_1/V_2 = 3.5$ $(\times 1.35)$; c – way C $(+l)$; $V_1/V_2 = 10(\times 0.45)$; d – way $A(+l)$; $V_1/V_2 = 10(\times 0.45)$.

The B type is analogous to the local "torch" regime (Fig. 15 c–f) with the broadening of the reaction front boundary as the point of the supply of reagent 1 is moved away along the flow axis (Fig. 50).

Type A—the planar reaction front—is a structure, perpendicular or at some other angle to the reactor axis and similar to that given in Fig. 15 (a, b). This is the regime of quasi-ideal displacement in very fast chemical reactions in turbulent flows; the coordinates of the planar reaction front along the reaction axis in this case may have both positive and negative values $(\pm l)$ relative to the point of reagent 1 supply (Fig. 50b, c). The latter results from back diffusion of the reagent flow.

Type Z is characterized by the absence of the reaction front. The reaction runs in the entire reaction volume, that being determined by strong back diffusion of the reagent flows with the tear flows away (Fig. 50d).

It is of great importance that not the absolute values of V_1 and V_2 is set their ratio V_1/V_2 plays a determining part in the formation of the particular front the reaction (B, A, Z) (Table 10).

The results of the experiment show that front B is manifested in the most distinct way only when the reagents are supplied by method $A(+l)$, i.e., using a feed pipe to supply reagents 1 with an open exit end at low (< 2) ratios of the flow speeds V_1/V_2, when sufficiently undeveloped flow turbulence is realized. The proceeding of fast chemical reactions

Table 10. The structure of the reaction front versus the methods of the reagent supply and the speeds ratio of the reagent supply

The way of the reagent supply	V_1/V_2	The type of the reaction front
A (+ l)	0.5–1.5	B
	1.5–2.0	mixed (B and A)
	2.0–5.0	A
	5.0–8.0	mixed (A and Z)
	$\geqslant 8.0$	Z
A ($l = 0$)	$\leqslant 1.5$	A
	1.5–2.0	mixed (A and Z)
	$\geqslant 2.0$	Z
A (− l)	0.5–25	Z
B (+ l)	0.5–60	A
	$\leqslant 60$	Z
B ($l = 0$)	0.5–60	Z
B (− l)	0.5–60	A

under such conditions proves inefficient due to the formation of the reagent slip zones, as well as due to the gradient of the reagent concentrations and temperatures along the reactor coordinates.

Front Z is formed in all methods of supplying reagents at relatively high ratios of V_1/V_2, as compared with the formation of fronts B and A. For instance, the reaction zone Z in the whole reaction volume, determined by the back diffusion of reagent flows 1 and 2, is realized by supplying the reagents A (+ l) at $V_1/V_2 \geqslant 8$ and B (+ l) – at $V_1/V_2 = 60$. It should be borne in mind that in supplying the reagents according to methods A and B at $l = 0$, i.e., in the reagent supply at a single point, type Z appears to be the only macroscopic low efficiency method of reagent supply. Due to the separated flows, the stagnation zone of the length (l_1) is formed in the tubular reactor. It is evident that in very fast chemical reactions in the stream turbulent reactors under regime Z, the time that a part of the product stays in the reaction zone increases. Naturally, this may lead to the formation of various secondary processes.

The planar front A (the regime of quasi-ideal displacement in turbulent flows) proves to be the most favorable in very fast chemical reactions. It represents an intermediate point between regimes B and Z, due to sufficiently well-developed turbulence in the mixing zone of reagents 1 and 2. Front A is characterized by uniformity of reaction flow composition along the reaction zone radius.

At the radial method of the reagent supply (the way of supplying the reagents B(+ l)), the structure of reaction front A appears to be much more stable than for method A(+ l). The planar reaction front is formed at the wide interval of values of V_1/V_2. However, in this case the boundary of the mixing front always has a negative corrdinate along the reactor axis in relation to the point of supplying the reagent 1: the reaction front shifts due to back diffusion. The higher the numerical value of V_1/V_2, the further the boundary of the planar reaction front shifts to the side opposite to the flow direction in the reactor.

Only in mixing two turbulent flows of reagents 1 and 2 according to methods A(+l) and B (−l), the planar reaction front A is formed at a certain strictly definite distance h from the point of supplying reagent 1 with the positive coordinate along the reactor axis. Here, method A(+ l) is in agreement with the empiric equation h/d = 1–0.2·V_1/V_2 with a correlation factor of 0.93. Thus, in supplying the reagents 1 and 2 according to the methods A(+ l) and B (− l), a peculiar "dead" zone occurs, followed by spasmodic (rases) formation of front A.

It should be noted that when the ratio of V_1/V_2 increases, in method A(−l) front A shifts in the direction of the flow in the reactor—a phenomenon inverse to that observed with methods A(+l) and B(+l).

Mathematical modeling has been conducted of mixing by method A(+1) within "q–ε" model of turbulence, taking into account the hydrodynamic conditions' effect on the mixing and on the chemical transformation, to study the process of the formation of planar reaction front A in very fast chemical reaction turbulent flows. These equations for the average values of pulsating quantities have been used:

$$\partial(\rho V_j)/\partial x_j = 0 \tag{39}$$

$$\partial(\rho V_j V_i)/\partial x_j - \partial(\tau_{ij})/\partial x_j + \partial p/\partial x_i = 0 \tag{40}$$

$$\partial(\rho V_j H)/\partial x_j + \partial(JH_j) = 0 \tag{41}$$

$$\partial(\rho V_j Y_s)/\partial x_j + \partial(J_{sj})/\partial x_j - \sum_{r=1}^{N_R} W_{sr} = 0 \tag{42}$$

where $S = 1, 2 \cdots N_R$; ρ, p, H are the density, pressure and enthalpy of the mixture stagnation respectively; Y_s is the mass concentration of the S-th component, N_R is the number of reactions and components; J_J—the j-th components of the vectors of mass or heat flows densities; τ_{ij}—the viscous stress tensor (considering the Reynolds stresses). To close the system of equations (43)–(46) using the averaged values, added by the equation of state $\rho = \rho(T)$, we have used a half-empiric two-parameters model of turbulence that contained two additional equations—that of kinetic energy of the pulsation movement q and that of the average specific dissipation rate ε. The speeds of flows 1 and 2 and the reactor size corresponded to the experimental data.

Some results of the computation are summarized in Fig. 51. The turbulent flow energy at the outflow of reagent has its maximum in the

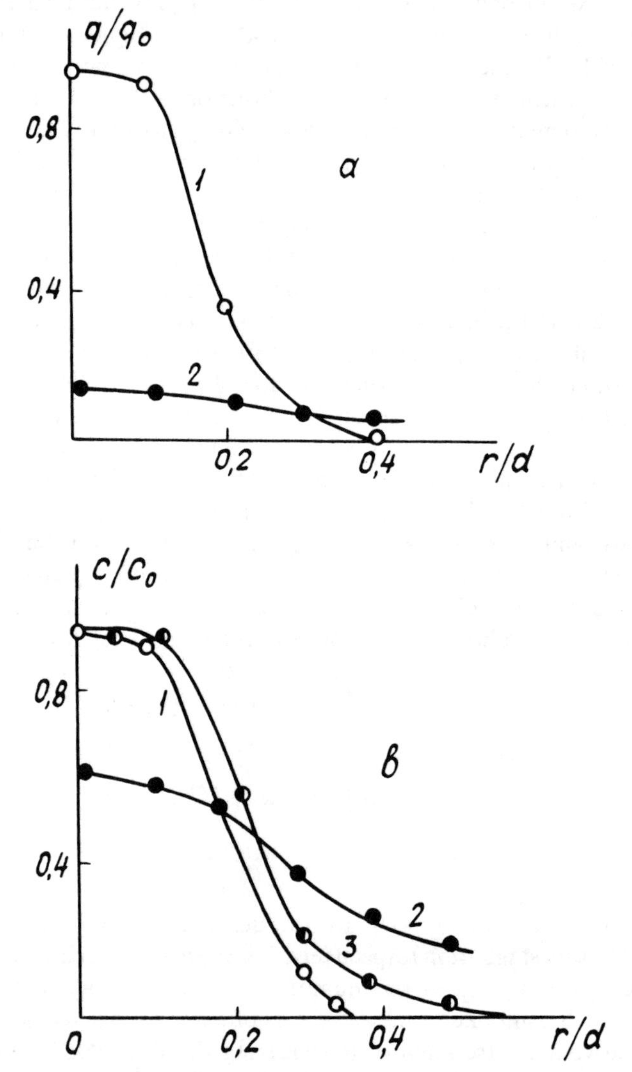

Figure 51. The distribution along the reactor radius of: a – the flow turbulent energy at $V_1/V_2 = 1.63$; z/d: 1 – 3.1; 2 – 4.0; b – reagent I (mass portion) at 1 – $V_1/V_2 = 1.0$; $z/d = 3.1$; 2 – $V_1/V_2 = 1.0$; $z/d = 4$; 3 – $V_1/V_2 = 2.2$; $z/d = 3.1$.

center of the flow, and as it moves away from the point of the outflow of reagent 1 both absolute value of the turbulent energy and its gradient along the reactor radius considerably decrease. At the distance of $Z/d = 3$ from the point of the outflow of reagent 1, the jet still retains its nucleus; however, at $Z/d > 3$, the jet breaks down and a practically uniform (mixed) flow is observed (Fig. 51 a, b).

At the low values of $V_1/V_2 < 1.5$, the reaction front B is realized, i.e., the "torch" with the occurrence of the zone where the reagent flow does not penetrate. Consequently, a chemical reaction does not proceed inthe vicinity of the reactor walls. The increase of V_1/V_2 results in the appearance of reagent 1 in the wall region and thus in marked reduction of the gradient of the reagent concentrations along the reactor radius (Fig. 51a, b). If one takes into account the experiment conditions this visually corresponds to the formation of front A, that is a significant fact (Table 10).

5 OPTIMIZATION OF MOLECULAR-MASS CHARACTERISTICS OF A POLYMER IN FAST POLYMERIZATION PROCESSES

5.1 THE POLYMER MASS-MOLECULAR CHARACTERISTICS AT THE MULTI-STAGE CATALYST SUPPLY IN THE REACTING TURBULENT FLOW WITHOUT HEAT REMOVAL

In polymerization processes under the conditions of adiabatic heating of reaction mass, the problem is the increase of the average MW and narrowing of the MWD, as high absolute values of temperature lead to the production of low-molecular products with high polydispersity, i.e. with the wide MWD [2, 58]. In this connection the solution of the problem of obtaining polymer products with the maximum values of $\bar{P}_w$, $\bar{P}_n$ and $\bar{P}_z$ at the most narrow MWD is of great importance.

As was mentioned above, these are the determining factors in fast liquid-phase polymerization in a tubular reactor with the turbulent flow under the conditions of quasi-ideal displacement (the macroscopic regime A [31, 32]; see Section 2.1, Chapt. 2): the geometry of the reaction zone, the initial monomer concentration, the original temperature of the monomer solution (the amount of the supplied catalyst is not great and its temperature exerts no effect on the system temperature at the reactor inlet), the speed of the movement reacting flow and its turbulization at the point of mixing of the solutions of the catalyst and monomer (in the active zone), the total concentration of the catalyst, depth of the

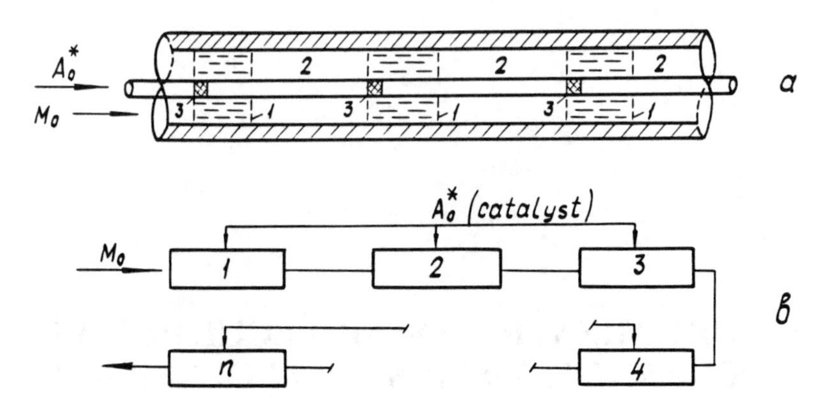

Figure 52. The diagram of the multi-stage catalyst supply: a – the zone Model; b – the cascade of reactors.

monomer conversion, the boiling temperature which depends on the chemical nature of the solvent and pressure, the correlation of the coefficients of heat and mass transfer as well as the equipment that makes it possible to create the anisotropic mechanism of heat transfer [23, 29–31, 37–42, 59–61].

We have studied the possibilities of controlling the polymer MW and MWD by variation of the catalyst supply, in particular, at the catalyst supply by stages that may be achieved both using the zone model of the reactor (Fig. 52a) and applying a cascade of tubular reactors with the catalyst supplying into each reactor (Fig. 52b) [39, 42, 62].

The zone model of a reactor is a consecutive connection of a few independent reaction zones where the reaction mixture comes after the process in the preceding zone is completed and where a new portion of the catalyst solution or that of the catalyst and monomer is supplied (Fig. 52a). In each zone, the temperature is constant and determined by the heat balance with take into (or without it) boiling and heat transfer through the outer wall.

Such a model is similar to a cascade of reactors of ideal displacement in each of which the constant temperature is maintained. Correspondingly, the conclusions are applicable to a cascade of series—connected (consecutive) reactors of ideal displacement or mixing, with the certain conditions imposed on the temperature in each reactor according to the selected variant of thermal balance.

For the realization of the zone model, the condition of the uniformity of the temperature and concentration fields in the turbulent conditions of the stream reacting flow with $R_{cr} \leqslant (D_T/k_t)^{1/2}$ should be met [59] (see section 2.1, Chapt. 2); restricting at the top $k_t = 17.5\,s^{-1}$ at 263 K [32, 35].

The intensity of heat evolution in each zone is dependent on the polymerization rate, provided that the amount of reacting monomer is not high and heat exchange coefficients would support the assumption that the temperature is constant in each zone, i.e., in the reaction zone a polymer with the exponential MW and MWD, corresponding to the temperature determined by the thermal balance, is formed. Calculation has shown that similar conditions are satisfied at

$$\Delta T_i \cong \frac{\delta \cdot V \cdot Q_{11} \Delta M_i}{\lambda_T} \ll \Delta T_2 = \frac{R \cdot T_i^2}{E_m - E_p} \tag{43}$$

where ΔT_i is the characteristic temperature gradient in the i-th reaction zone; $V \cdot Q_{11} \Delta M_i$ is the rate of the heat evolution in the i-th zone; V is the linear flow rate; Q_{11} is the polymerization heat in the condensed phase; ΔM_i is the amount of the monomer that has reacted in the i-th zone; $\frac{\lambda_T \Delta T_i}{\delta}$ is the rate of "spreading" of heat along the reactor; λ_T is the effective coefficient of heat conductivity; $\delta = V/k_t$ is the characteristic size of the reaction zone; ΔT_2 is the characteristic interval of temperature for the MW variation; E_m, E_p are the activation energies of the chain transfer on monomer and chain propagation reactions; T_i is the average temperature in the i-th zone; R is gas constant.

The reaction zones do not overlap when the distance between the neighboring points of the catalyst supply satisfies the criterion $l_i \gg \delta = V/k_t$, when adiabatic heating of the reaction system is considered.

In case the heat evolved in polymerization heats the reaction mass and there is an unambiguous dependence of the amount of the monomer that has reacted in all preceding zones including the i-th one on the average temperature of the i-th zone T_i and on the initial temperature T_0, then:

$$T_i = T_0 + \alpha \sum_{k=1}^{i} \Delta M_k \tag{44}$$

where $\alpha = Q_{11}/C_p$ (C_p is the average specific heat of the reaction mass).

Since the polymer is formed independently in each zone, the total MWD is a sum of the MWD functions in each zone:

$$\rho_w^i(j) = \sum_{k=1}^{i} \Delta M_k \rho_w^k(j) / \sum_{k=1}^{i} \Delta M_k \tag{45}$$

Here $\rho^k_w(j)$ is weight function of the MWD of the polymer obtained in each zone; $\rho^i_w(j)$ is the weight function of the MWD of the whole amount of the polymer after the i-th zone.

Correspondingly, expressions for the average degrees of polymerization are as follows:

$$\bar{P}^i_n = \sum_{k=1}^{i} \Delta M_k / \sum_{k=1}^{i} \Delta M_k / P^k_n \tag{46}$$

In case the cascade of reactors consists of n apparatus, the equations determining the average values of MW have the following form:

$$\bar{P}^i_w = \sum_{k=1}^{i} \Delta M_k P^k_w / \sum_{k=1}^{i} \Delta M_k \tag{47}$$

$$\bar{P}^i_z \bar{P}^i_w \sum_{k=1}^{i} \Delta M_k = \sum_{k=1}^{i} P^k_z P^k_w \Delta M_k \tag{48}$$

Equations (46)–(48) are true irrespective of the distribution of the polymer in each zone.

For further analysis, the whole kinetic scheme and the polymerization mechanism are not important. Only the fact that the average MW and MWD are determined by the reaction of the chain transfer to monomer is essential. Taking into account the practically constant reaction temperature in each zone, we obtain:

$$\rho^k_w(j) = \frac{j}{(\bar{P}^k_n)^2} e^{-j/P^k_n}$$

$$\frac{1}{3} P^k_z = \frac{1}{2} P^k_w = P^k_n = \frac{k^0_p}{k^0_m} e^{(E_m - E_p)/RT_k} \tag{49}$$

where k^0_p, k^0_m, E_p, E_m are the frequency factors and activation energies at the stages of chain propagation and transfer on monomer.

Thus, we have a closed system of equations (44)–(49) enabling us to calculate the MWD and the average MW for the arbitrary distribution of the polymer yields in each zone (ΔM_i). To establish the relationship between the polymer yield in the given zone and the amount of the catalyst supplied, one should have more profound knowledge of the polymerization kinetics. In particular, using the scheme of cationic polymerization of isobutylene (VI)–(X) [1, 10, 11, 33], we obtain

$$\Delta M_i = \left(M_0 - \sum_{k=1}^{i-1} \Delta M_k\right)\cdot(1 - e^{-k_p/k_t A_i^*})$$

$$\sum_{k=1}^{i} \Delta M_k = M_0 \left[1 - \exp\left(-\sum_{k=1}^{i} \frac{k_p}{k_t} A_k^*\right)\right] \tag{50}$$

where k_p, k_t are the effective constants of the chain propagation and termination rates respectively in each zone: A_i^* is the amount of the catalyst supplied in the i-th zone, $M_0 -$ the concentration of the monomer in the initial mixture. If k_p and k_t are not dependent on the temperature, that, being characteristic of isobutylene polymerization in particular, [1, 10, 25, 33], Eq. (44) is simplified to:

$$\sum_{k=1}^{i} \Delta M_k = M_0 \left[1 - \exp\left(-\frac{k_p}{k_t} \sum_{k=1}^{i} A_k^*\right)\right] \tag{51}$$

Eqs. (44)–(49) make it possible to calculate the MW and the MWD of a polymer at any given method of the catalyst supply in polymerization under the adiabatic conditions, i.e., without heat removal.

5.2 THE TWO-ZONE REACTOR MODEL

For the two-zone reactor eq. (44) is transformed to:

$$\bar{P}_w = 2\frac{\Delta M_1}{\Delta M}\cdot\frac{k_p^0}{k_{tr}^0} \mathrm{e}^{-(E_p - E_{tr})/RT_1} + 2\frac{\Delta M_2}{\Delta M}\frac{k_p^0}{k_{tr}^0} e^{-(E_p - E_{tr})/RT_2} \tag{52}$$

Here and further on ΔM defines the total amount of the polymer at the outlet of the reactor, and

$$P_w^0 = \frac{k_p^0}{k_{tr}^0} e^{-(E_p - E_{tr})/T_0}$$

is the polymerization degree of the polymer obtained under the isothermal conditions at the temperature of the supplied monomer solution T_0.

After transformation, taking that $\dfrac{E_{tr} - E_p}{R} = \varepsilon$, we obtain:

$$\exp\left(\frac{\varepsilon}{T_1}\right) = \exp\left(\frac{\varepsilon}{T_0 + \alpha\Delta M_1}\right) = \exp\left[\frac{\varepsilon}{T_0}\left(1 - \frac{\alpha\Delta M_i}{T_0}\right)\right]$$

$$= \exp\frac{\varepsilon}{T_0}\exp\left(-\frac{\varepsilon\alpha\Delta M}{T_0^2}\right) \tag{53}$$

where $\alpha = q/C_r$; q is the polymerization heat; C_r is the average specific heat of the reaction mass. This equation is true when $\alpha\Delta M_1/T_0 \ll 1$. This condition is always met at the earlier stages of the process when ΔM_1 is not high ($\alpha\Delta M_i \cong 20$–$40°$ at $T_0 = 200$–250 K).

These stages in the fast polymerization processes are particularly significant, since the product of highest molecular mass is formed there. We should note that under real conditions the maximum temperature drop in the reaction zone does not exceed 100°, i.e. the maximum values of $\alpha\Delta M/T_0 \cong 0.5 < 1$. Therefore, despite the rough approximation, the obtained results are qualitatively and often quantitatively true.

Transforming Eq. (52) we obtain:

$$\bar{P}_w/P_w^0 = \frac{x_1}{x}e^{-x_1} + \frac{x_2}{x}e^{-x} \tag{54}$$

$$\bar{P}_n/P_n^0 = \frac{x_1}{x}e^{x_1} + \frac{x_2}{x}e^{x} \tag{55}$$

$$\bar{P}_Z\bar{P}_w/P_z^0P_w^0 = \frac{x_1}{x}e^{-2x_1} + \frac{x_2}{x}e^{-2x} \tag{56}$$

Here $x = \dfrac{\varepsilon\alpha\Delta M}{T_0^2}$, $x = x_1 + x_2$.

The parameter x at the end of the reaction (at the reactor outlet) considerably exceeds one ($x \gg 1$). This fact agrees with the above assumption since $\varepsilon/T_0 \gg 1$. Here and further on we use the assumption $x \gg 1$. This surely does not concern the values of the parameter x_1 at various intermediate stages.

The characteristic dependences of $\bar{P}_n$, $\bar{P}_w$ and $\bar{P}_z$ on x_1/x (the portions of the polymer obtained in the first zone) are given in Fig. 53a and the dependence of the MWD width ($\bar{P}_w/\bar{P}_n$ and $\bar{P}_z/\bar{P}_w$) on x_1/x is given in Fig. 53b. It is seen that the consecutive supply of the catalyst in each of the cascade reactors allows us to achieve the maximum value of this or that average MW. The decrease of the catalyst amount in the first reactor leads to the natural reduction of the reaction mass temperature, which in turn results in the increase of the mean MW of the polymer formed in this zone. However, the amount of the polymer produced in the first zone correspondingly decreases. This fact causes the reduction of the polymer effect contribution to the average MW. The competition of these factors results in the extremum dependence of the average MW on the portion of the catalyst supplied into the first reactor.

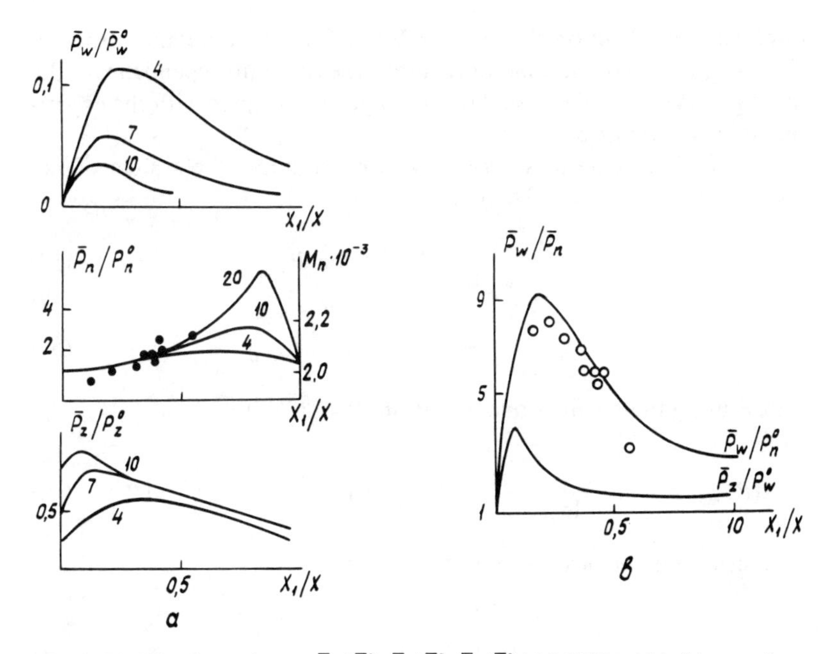

Figure 53. The dependences $\bar{P}_w/\bar{P}_w^0$, $\bar{P}_n/\bar{P}_n^0$, $\bar{P}_z/\bar{P}_z^0$ (a) MWD width (b) on x_1/x at $x = 10$. The figure near curves corresponds to x. Points on the curves reflect experimental data.

The two-zone way of the catalyst supply allows us to vary two parameters–the total amount of the catalyst (or the total yield of the polymer ΔM) and the distribution of the catalyst shares in two zones. The reduction of the total amount of the catalyst and, accordingly, of ΔM or x always leads to the rise of the average MW and narrowing of the MWD. The polymer yield and, consequently, the process efficiency are reduced. To conduct the process under the conditions close to the isothermal ones, the condition $x \ll 1$ must be satisfied, corresponding to the polymerization of the mixture of isobutylene with isobutane in the ratio of 1:1. Then $\Delta M \ll 7\%$wt (from isobutylene at $\varepsilon = 2000°$ and $\alpha = 600°$). Actual conditions require the process to be conducted up to the degree of 70 %wt and more. In this case $x > 10$.

Varying x_1/x (Fig. 53), one may obtain the maximum value of the average MW. Actually, the reduction of the catalyst portion in the first zone results in the temperature drop T_1 and the average mean MW increase of the polymer formed in the first zone, the polymer MW being constant at the given value $x = \text{const}$ in the second zone. But simultaneously the polymer amount in the first zone is reduced and accordingly diminishes its effect on the average MW of the final polymer product. The competition of these factors causes the extremum dependence of the average MW on x. The calculated data generally agree with the experimental results (Fig. 53).

To find extrema it is necessary to differentiate the system of Eq. (54)–(56) to x_1. At $x \gg 1$, we obtain:

$$x_1^{\max} \cong \begin{cases} x^{1/2}e^{-x/2} & \text{for } \bar{P}_{z\,\max} \quad (57) \\ 1 & \text{for } \bar{P}_{w\,\max} \quad (58) \\ x - \ln\, x & \text{for } \bar{P}_{n\,\max} \quad (59) \end{cases}$$

Substituting these values in correlations (54)–(56) it is:

$$(\bar{P}_n)_{\max} \cong P_n^0\, e^{-x}\frac{x}{\ln x + 1}; \quad (\bar{P}_w)_{\max} \cong P_w^0\frac{1}{e\cdot x}; \quad (\bar{P}_z)_{\max} \cong (P_z^0) \tag{60}$$

or considering the next order of accuracy:

$$x_1^{\max} = \begin{cases} x^{1/2}e^{-x/2}(1 - x^{1/2}/2e^{-x/2}) \\ 1 + e^{-(x+1)} \\ x - \ln\, (1 + x - \ln x) \end{cases} \tag{61}$$

$$\bar{P}_z/P_z^0 \cong 1 - 2x^{1/2}e^{-x/2}, \quad \bar{P}_w/P_w^0 \cong \frac{1}{ex}(1 + xe^{-(x-1)}), \tag{62}$$

$$\bar{P}_n/P_n^0 = e^{-x}\frac{x}{lnx + 1}\left(1 + \frac{2}{x}\right) \tag{63}$$

At the constant value of the total yield of the polymer ($x = \text{const}$), there remains only one variable parameter x_1/x; correspondingly the optimum value may be only obtained for some particular MW. For instance, the maximum $\bar{P}_w$ is reached when $x \cong 1$, close to the maximum MWD, is achieved at the same time ($\bar{P}_w/\bar{P}_n$).

If the value of $\bar{P}_w$ is not limited, increasing the catalyst supply in the first zone ($x_1 > 1$) may narrow the MWD on account of some reduction of $\bar{P}_w$.

A certain increase of $\bar{P}_n$ in ($x/ex + 1$) times may be achieved by supplying the catalyst in two steps ($x_1 \cong x - lnx$), as compared to a single, one-time supply of the catalyst. The MWD of the polymer formed in this case does not appear to be too wide, but at any rate wider than the exponential one ($\bar{P}_w/\bar{P}_n \cong 2\cdot\ln x$) and $P_z/P_w \cong 3/2\cdot\left(1 + \frac{\ln x}{x}\right)$. The maximum $\bar{P}_z$ is extremely flat towards the increase of x_1, the drop of $\bar{P}_z$ after the flex point corresponds to the law $\bar{P}_z/P_z^0 = e^{-x_1}$ (at $x_1 = x^{1/2}e^{-x/2}$). Hence, comparatively high values of $\bar{P}_n$ may be obtained at $x = 1$ as well.

5.3 THE THREE-STAGE CATALYST SUPPLY

The catalyst influx in the reacting system by three steps gives great opportunities to regulate, control and optimize the MWD since there appears another independent governing parameter–the amount of the catalyst supplied into the second zone.

First, the three-zone reactor allows us to obtain higher values of the mean MW when tending to obtain the maximum value for one of them. In this case, expressions for $\bar{P}_w$ and $\bar{P}_n$ can be written as follows (using equations (55) and (60));

$$\bar{P}_w/P_w^0 = \frac{x_1}{x}e^{-x_1} + \frac{x_2}{x}e^{-(x_1+x_2)} + \frac{x - x_1 - x_2}{x}e^{-x} \tag{64}$$

$$\bar{P}_n/P_n^0 = \frac{x_1}{x}e^{x_1} + \frac{x_2}{x}e^{(x_1+x_2)} + \frac{x - x_1 - x_2}{x}e^{x} \tag{65}$$

Differentiating expression (55) to x_1 and x_2 and making the partial derivatives $\partial\bar{P}_w/\partial x_1$ and $\partial\bar{P}_w/\partial x_2$ equal to zero, we obtain the absolute minimum. At $x \gg 1$ we have $x_1 \cong 1 - \frac{1}{e}$, $x_2 \cong 1$. Substituting the values of x_1 and x_2 in Eq. (64) we obtain

$$(\bar{P}_w/P_w^0)_{\max} \cong \frac{1}{e\cdot x}e^{1/e} \cong \frac{1.5}{e\cdot x} \tag{66}$$

that being 1.5 times more than the maximum possible value of $\bar{P}_w$ in the two-stage catalyst supply.

Differentiating Eq. (65) to x_1 and x_2 and making the partial derivatives equal to zero at $x \gg 1$, we obtain:

$$\left(\frac{\bar{P}_n}{P_n^0}\right)_{\max} \cong e^{-x}\frac{x}{1 + \ln(1 + \ln x)} \tag{67}$$

This value is also higher than for the two-zone reactor. It is obtained at $x_1 \cong x - \ln(x - \ln x) - \ln(1 + \ln x)$; $x_2 \cong \ln(x - \ln x)$ and $x_3 \cong x - x_1 - x_2 \cong \ln(1 + \ln x)$. The three-stage catalyst supply offers other opportunities for the MWD optimization. For instance, one can obtain close to the maximum values of two average MW simulataneously, separately optimizing $\bar{P}_w$ to x_1 and $\bar{P}_n$ to x_2. If we know the location of maxima (Fig. 36) in the region of the maximum $\bar{P}_w$, we can neglect the two last terms in expression (55), and in the region of the maximum $\bar{P}_n$—the first term in Eq. (60).

Differentiating first $\bar{P}_w$ to x_1 and then—$\bar{P}_n$ to x_2 at $x = \text{const}$ and making the corresponding derivatives equal to zero we obtain:

$$x_1 \cong 1, x_2 \cong x - 1 - \ln x.$$

Substituting these values in Eq. (64) and (65) we have:

$$(\bar{P}_w)_{\max} \cong \frac{P_w^0}{e \cdot x}; \quad (\bar{P}_n)_{\max} \cong P_n^0 e^{-x} \frac{x}{\ln x + 1} \tag{68}$$

Eq. (68) practically coincides with Eq. (60). However, the fact that values of $\bar{P}_w$ and $\bar{P}_n$ are obtained simultaneously due to the possibility of three places of the catalyst supply is essential too. In a general case, the dependence of the MW on x_1/x and x_2/x (Fig. 54) represents a surface. Knowing the profile of the variation surface of the MW due to the

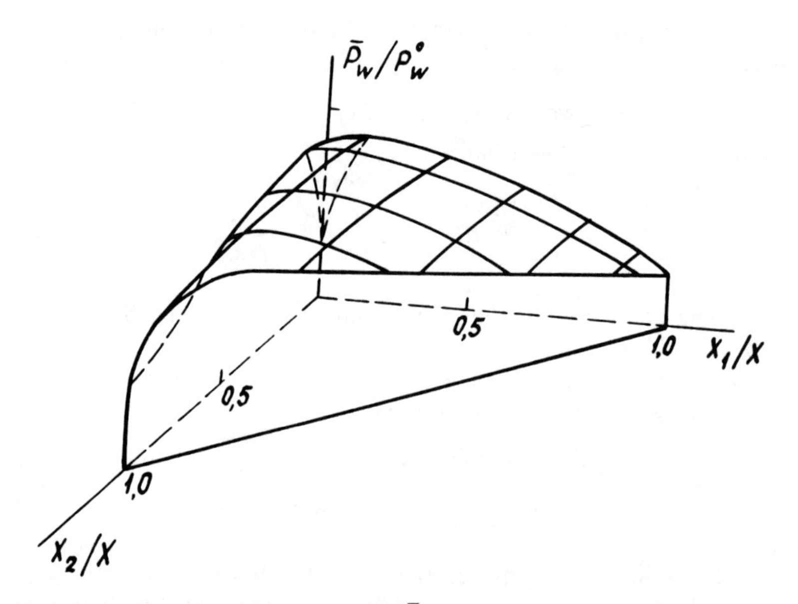

Figure 54. The surface of dependence of $\bar{P}_w$ on the catalyst portion supplied into the first x_1/x and second x_2/x reactors of cascade.

parameter x, one may choose close to optimum values of the concentrations of the catalyst supply into each zone, to obtain the polymer with the given MW and MWD, i.e., to effectively influence the MW and MWD of the polymer products formed. The absolute maximum $\bar{P}_w$ is reached at $x_1 \cong 1 - 1/e$ and $x_2 \cong 1$. The sufficiently narrow MWD is produced near the maximum $\bar{P}_n$ or in the case that all the catalyst is supplied in the first zone.

5.4 THE FOUR-STAGE CATALYST SUPPLY

The use of four places of the catalyst supply (four zones of the reaction) allows us at the constant x to increase the average MW values to a still greater extent or the optimize the MWD at the same time. As an example, we give the results of the determination x_1,x_2 and x_3 with the aim of simultaneously obtaining the maximum values of $\bar{P}_w$,$\bar{P}_n$ and $\bar{P}_z$. Under the condition that $x \gg 1$, as was shown at the two-stage catalyst supply, the maximum values of $\bar{P}_w$, $\bar{P}_n$ and $\bar{P}_z$ considerably differ from each other and may neglect the respective terms in Eq. (54)–(56) at their optimization. Therefore, the values of the optimum parameters x, x_1 and x_2 in fact coincide with the corresponding values of the two-zone reactor. The values of $(\bar{P}_z)_{max}$, $(\bar{P}_w)_{max}$ and $(\bar{P}_n)_{max}$ coincide too (Table 11).

The difference is that in two-zone reactor it is possible to realize only one of the variants and reach the maximum values of either $\bar{P}_w$ or $\bar{P}_n$ or $\bar{P}_z$, while in the four-zone reactor it is possible to achieve the same values of the mean MW simultaneously if necessary (Tables 11, 12).

5.5 THE MULTI-STAGE CATALYST SUPPLY

There exists another limiting case of the catalyst supply into the reaction zone, where there are many reactors and very little amount of the catalyst is supplied into each of them. This is the case of continuous catalyst supply in a rather long reaction zone. From the kinetic aspect, this model is identical to the reactor of ideal displacement or to a batch reactor with the arbitrary catalyst supply into the reaction zone. In this case the expressions for the mean MW are as follows:

$$\bar{P}_w = \frac{1}{\Delta M}\int_0^{\Delta M} P_w\, dM = \frac{1}{\Delta M}\int_0^{\Delta M} 2\frac{k_p^0}{k_{tr}^0}\exp\left(\frac{\varepsilon}{T_0 + \alpha\Delta M}\right) dM \tag{69}$$

$$\bar{P}_n = \Delta M\left[\int_0^{\Delta M}\frac{dM}{P_n}\right]^{-1} = \Delta M\left\{\int_0^{\Delta M}\frac{k_{tr}^0}{k_p^0}\exp\left(-\frac{\varepsilon}{T_0 + \alpha\Delta M}\right) dM\right\}^{-1} \tag{70}$$

Table 11. The formulas for computation the optimum mass-molecular parameters of the MWD for the cascade of tubular adiabatic reactors at different ways of independent catalyst supply

The number of zones	1	2	2	2	3
MW	–	$\bar{P}_{n\max}$	$\bar{P}_{w\max}$	$\bar{P}_{z\max}$	$\bar{P}_{w\max}$ $\bar{P}_{n\max}$
$\bar{P}_w/P_w^0$	* e^{-x}	* $\sim xe^{-x}(1-2\ln x/x)$	* $(1+xe^{-(x+1)})/ex$	* $\sim(e^{-x/2}/x^{1/2})\cdot(1-x^{1/2}e^{x/2}/2)$	$\sim 1/(ex)$
$\bar{P}_n/P_n^0$	* e^{-x}	* $xe^{-x}(1+2/x)/(\ln x+1)$	* $e^{-x}(1+1/x)$	* $\sim e^{-x}\cdot(1-e^{-x/2}/x^{1/2})$	$e^{-x}\cdot x/(\ln x+1)$
$\bar{P}_z/P_z^0$	* e^{-x}	* $\sim xe^{-x}(1-\ln x/x)$	* $(1-x\cdot e^{-(x-1)})/e$	* $1-2x^{1/2}e^{-x/2}$	$\sim 1/e$
$\bar{P}_w/\bar{P}_n$	* 2	$\sim 2(\ln x+1)$	$\sim 2e^{x-1}/x$	$\sim 2e^{x/2}/x^{1/2}$	$\sim 2e^{x-1}\cdot(\ln x+1)/x^2$
$\bar{P}_z/\bar{P}_w$	* $3/2$	$3(1+\ln x/x)/2$	$\sim 3x/2$	$\sim 3x^{1/2}e^{x/2}/2$	$\sim 3x/2$
x_1	x	* $x-\ln(1+x-\ln x)$	* $\sim 1+e^{-(x+1)}$	* $x^{1/2}e^{-x/2}\cdot(1-x^{1/2}e^{-x/2}/2)$	~ 1
x_2	–	$\ln(1+x+\ln x)$	$\sim x-1$	$x-x^{1/2}e^{-x/2}$	$x-\ln(1+x-\ln x)-1$
x_3	–	–	–	–	$\ln(1+x-\ln x)$
x_4	–	–	–	–	

MW	$\bar{P}_{w\,max}$	$\bar{P}_{n\,max}$	$\bar{P}_{n\,max}, \bar{P}_{n\,max}, P_{w\,max}$	–
$\bar{P}_w/P_w^0$	$\sim e^{1/e}/ex$	$\sim xe^{-x}(1+\ln x)\cdot(1-2\ln x/x)$	$\sim 1/ex$	* $(1-e^{-x})/x$
$\bar{P}_n/P_n^0$	$\sim e^{-x}$	$\sim xe^{-x}/(1+\ln(1+\ln x)$	$xe^{-x}/(1+\ln x)$	* $x(1+e^{-x})/e^x$
$\bar{P}_z/P_z^0$	$\sim(1+1/2e^2)/e$	$\sim xe^{-x}(1-\ln x/x)\cdot(1+\ln x)$	~ 1	* $(1+e^{-x})/2$
$\bar{P}_w/\bar{P}_n$	$2e^{x-1}\cdot e^{1/e}/x$	$\sim 2(1+\ln x)\cdot[1+\ln(1+x)$ $(1-2\ln x/x)]$	$\sim 2e^{x-1}(1+\ln x)/x^2$	$2e^x(1-2e^{-x})/x^2$
$\bar{P}_z/\bar{P}_w$	$\sim 3x/2e^{1/e} \cong x$	$\sim 3(1+\ln x/x)/2$	$\sim 3/(2ex)$	* $3(1+2e^{-x})/4x$
x_1	$1-1/e$	$x-\ln(x-\ln x)-\ln(1+\ln x)$	$\sim x^{1/2}e^{-x/2}$	$\ll x^{1/2}e^{-x/2}$
x_2	1	$\ln(x-\ln x)$	~ 1	–
x_3	$x-2+1/e$	$\ln(1+\ln x)$	$x-1-\ln(1+x-\ln x)$	$\ll 1$
x_4	–	–	$\ln(1+x-\ln x)$	$\ll \ln x$

* The second term in brackets shows the order of accuracy of the conducted estimate $x=\alpha\varepsilon\Delta M/T_0^2=\varepsilon(T-T_0)T_0^2$, $(T-T_0)\ll 1$, $x\gg 1$.

Table 12. The calculated mass-molecular parameters of isobutylene polymerization for a cascade of the reactors with the independent catalyst supply

Parameter	The number of zones	$\bar{P}_n/\bar{P}_n^0 \cdot 10^{-4}$	$\bar{P}_w/\bar{P}_w^0 \cdot 10^2$	$\bar{P}_z/\bar{P}_z^0 \cdot 10^2$	$\bar{P}_w/\bar{P}_n \cdot 10^{-2}$	$\bar{P}_z/\bar{P}_w$	x_1	x_2	x_3	x_4
–	1	0.5	0.005	0.005	0.020	1.5	10	–	–	–
$\bar{P}_{n\,max}$	2	1.5	~0.040	~0.005	~0.045	1.8	7.5	2.3	–	–
$\bar{P}_{w\,max}$	2	0.5	3.500	35.00	14.00	15.0	1.0	9	–	–
$\bar{P}_{z\,max}$	2	0.5	0.220	100.0	0.900	700.0	0.02	20	–	–
$\bar{P}_{w\,max}$ $\bar{P}_{n\,max}$	3	1.5	~44.00	35.00	5.000	13.0	1	6.7	2.3	–
$\bar{P}_{w\,max}$	3	0.5	4.200	~35.00	21.00	10.0	0.65	1	–	–
$\bar{P}_{n\,max}$	3	~2.5	0.001	0.160	~0.100	1.80	~7	~2	~1	–
$\bar{P}_{z\,max}$, $\bar{P}_{w\,max}$, $\bar{P}_{n\,max}$	4	1.5	~4.000	100.0	5.000	~40.0	0.02	1	~6.7	2.3
–	∞	5.0	10.00	50.00	4.000	7.50	10^{-3}	10^{-3}	10^{-3}	10^{-3}

$$\bar{P}_z\bar{P}_w\,\Delta M = \int_0^{\Delta M} P_z P_w\,dM = \int_0^{\Delta M} 6\left(\frac{k_p^0}{k_{tr}^0}\right)^2 \exp\left(\frac{2\varepsilon}{T_0+\alpha\Delta M}\right)dM \quad (71)$$

At $\alpha\Delta M/T_0 \ll 1$ obtain

$$\bar{P}_w = P_w^0\frac{1-e^{-x}}{x} \cong \frac{P_w^0}{x};\ \bar{P}_n = P_n^0\frac{x}{e^x-1} \cong P_n^0 xe^{-x};$$

$$\bar{P}_z = 1/2P_z^0(1-e^{-2x})/(1-e^{-x}) \cong 1/2P_z^0(1+e^{-x})$$

or

$$\bar{P}_w = (P_w^0/x)\cdot(1-e^{-x});\ \bar{P}_n = P_n^0\cdot e^{-x}\cdot x\cdot(1+e^{-x});\ \bar{P}_z = 1/2P_z^0(1+e^{-x}) \quad (72)$$

The above process proceeding allows us to reach the maximum values of all moments of distribution. However, this does not correspond to the absolute maximum values of the average MW (the moment ratios), as $\bar{P}_w$ and $\bar{P}_n$ turn out to be higher and $\bar{P}_z$—lower than in a two-reactor block. All the calculated results are summarized in Table 11. In Table 12 gives the numerical values of the respective values for $x = 10$. The first reactors are seen to be most responsible for the formation of high-molecular products. Reducing the amount of the catalyst supplied in the first sections of the cascade of the reactors, one may increase the corresponding moments of distribution. Thus, in order to obtain the limiting values of $\bar{P}_z$, the initial concentration of the catalyst should meet the condition $x < x^{1/2}e^{-x/2}$. Then, obtaining $\sum x_i = 2 \div 3\cdot x^{1/2}\, e^{-x/2}$, the amount of the catalyst may be increased and must satisfy the condition $x_i \ll 1$. After $\sum x_i = 2 \div 3$, more catalyst may be added until $x_1 \ll \ln x$.

Thus, the above analysis has shown that due to the conditions of polymerization, to the number of the reactors in the cascade and to the catalyst volume in each of them in particular, one may intensely influence the molecular-mass characteristics (MWD, $\bar{P}_n$, $\bar{P}_w$, $\bar{P}_z$) of the polymer formed.

5.6 MASS-MOLECULAR CHARACTERISTICS OF THE POLYMER FORMED UNDER THE CONDITIONS OF INTERNAL HEAT REMOVAL

In many cases the solvent and/(or) the monomer starts boiling at a certain temperature. The boiling of the solvent or monomer allows us to markedly improve the polymerization conditions and to obtain higher values for the average MW and narrow MWD [40].

If the boiling temperature is reached in the last zone, the further fragmentation of zones, naturally, makes sense, as the reaction proceeds at the constant temperature. A single (one time) catalyst supply in this case may lead to the most narrow (exponential) MWD and the lowest mean MW:

$$1/3\bar{P}_z = 1/2\bar{P}_w = \bar{P}_n = k_p^0/k_{tr}^0 e^{-\varepsilon/T_k} = P_n^0 e^{-x_k} \tag{73}$$

where $x_k = \dfrac{\varepsilon(T_k - T_0)}{T_0^2}$

5.7 THE TWO-STAGE CATALYST SUPPLY UNDER THE CONDITIONS OF INTERNAL HEAT REMOVAL

The two-stage catalyst supply allows us to increase the average MW according to Eqs. (54)–(56) (Section 6.2, Chapt. 6).

The plots of the functions $\bar{P}_w$, $\bar{P}_n$ and $\bar{P}_z$ from (x_1/x) are analogous to the ones given in Fig. 53. Similarly, the values of x_1 leading to the maxima of the average MW can be found using Eqs (57)–(60).

In particular

$$(\bar{P}_w)_{\text{mas}} = \frac{P_w^0}{ex}(1 + xe^{-(x_k - 1)} \tag{74}$$

$$(\bar{P}_n)_{\max} = P_n^0\, e^{-x_k} \frac{x}{x - x_k + \ln x_k}\left(\frac{1}{x - x_k + \ln x_k}\right) \tag{75}$$

$$(\bar{P}_z)_{\max} = P_z^0\,(1 - 2x\, e^{-x_k/2}) \tag{76}$$

It should be noted that the second term in the brackets shows the order or accuracy which the estimated parameter values are given with.

It appears that the maximum values $\bar{P}_w$ and $\bar{P}_z$ are close to the ones obtained in the polymerization reaction without boiling and number-average MW is surely much higher then without boiling (e^{x_k} instead of e^{-x}). The two-stage catalyst supply during the boiling of the solvent and/(or) monomer does not produce such great a gain in $\bar{P}_n$ in comparison with the single-stage (one time) catalyst supply as in the reactor without boiling. The most prominent effect of boiling is observed in MWD narrowing.

$$\left(\frac{\bar{P}_w}{P_n}\right)_{\max} = 2\frac{e^{(x_k - 1)}}{x} \ll 2\frac{e^{(x - 1)}}{x} \tag{77}$$

This fact is evident from Tables 13 and 14 at $x = 10$ and $x = 5$. Correspondingly still more narrow MWD can be obtained if the production of the polymer with the maximum $\bar{P}_w$ is not desired. Thus, in producing the polymer with the maximum $\bar{P}_n$ the width of distribution is equal to:

$$\left(\frac{\bar{P}_w}{P_n}\right)_{P_n^{\max}} \cong 2\,\frac{x_k^2(x - x_k + \ln x_k)}{x^2} \tag{78}$$

when $(\bar{P}_n)_{\max}$

$$\frac{\bar{P}_w}{P_w^0} \cong \frac{x_k^2}{x_k} e^{-x_k}\left(1 - \frac{2\ln x_k}{x_k} + \frac{x - x_k}{x_k^2}\right) \tag{79}$$

$$\frac{\bar{P}_w}{\bar{P}_n} \cong 2\,\frac{x_k^2(x - x_k + \ln x_k)}{x^2}\left(1 - \frac{2\ln x_k}{x_k} + \frac{x - x_k}{x_k^2} + \frac{1}{x - x_k + \ln x_k}\right) \tag{80}$$

at $x - x_k \ll x_k^2 \ll e^{x_k}$, $x - x_k + \ln x \gg 1$

$$\frac{\bar{P}_z}{P_z^0} = x_k e^{-x_k}\left(1 - \frac{3\ln x_k}{x_k} + \frac{x - x_k}{x_k^3}\right) \tag{81}$$

at $x - x_k \ll x_k^3 \ll e^{x_k}$

$$\frac{\bar{P}_z}{\bar{P}_w} = \frac{3}{2}\frac{x}{x_k}\left(1 - \frac{\ln x_k}{x_k} - \frac{x - x_k}{x_k^2}\right) \tag{82}$$

that already may be very close to the width of the exponential MWD (Table 13).

5.8 THE MULTI-STAGE CATALYST SUPPLY IN TUBULAR REACTOR WITH INNER HEAT REMOVAL

An increased possibility to optimize the MWD is presented by a three-zone reactor with the boiling in the third zone. The average MW in this case is a function of two variable parameters (x_1/x and x_2/x)

$$\frac{\bar{P}_w}{P_w^0} = \frac{x_1}{x} e^{-x_1} + \frac{x_2}{x} e^{-(x_1 + x_2)} + \frac{x - x_1 - x_2}{x} e^{-x_k} \tag{83}$$

$$\frac{\bar{P}_n}{P_n^0} = \frac{x_1}{x} e^{x_1} + \frac{x_2}{x} e^{(x_1+x_2)} + \frac{x - x_1 - x_2}{x} e^{x_k} \tag{84}$$

$$\frac{\bar{P}_z \bar{P}_w}{P_z^0 P_w^0} = \frac{x_1}{x} e^{-2x_1} + \frac{x_2}{x} e^{-2(x_2-x_1)} + \frac{x - x_1 - x_2}{\mathrm{x}} e^{-2x_k} \tag{85}$$

It is important that absolute values of $\bar{P}_n$ should be essentially higher in the reactor with boiling and the MWD should be narrower. In the given case, the maximums of $\bar{P}_n$ and $\bar{P}_w$ are obtained at practically the same comparatively low values of x_1 or $(x_1 + x_2)$ as under the conditions without boiling Eqs. (57) and (58). The maximum value of the number-average MW is obtained with $x_1 + x_2 \cong x_k - \ln x_k$.

Under such operating conditions, for instance, one can obtain simultaneously the maximum values of $\bar{P}_z$ and $\bar{P}_w$ at $x_1 = x^{1/2} e^{-x_k/2}$ and $x_2 \cong 1$, or $\bar{P}_w$ and $\bar{P}_z$ at $x_1 \cong 1$, $x_2 = x_k - \mathrm{in}\, x_k$. The corresponding values of the maximum average MW are close to those obtained from Eq. (61) (Tables 13 and 14). Surely, in this case of polymerization without boiling, at three-stage catalyst supply the absolute maximum of the average MW, exceeding the maximum values, obtained in two-zone reactors, can be achieved. Since the limits of variation of the number-average MW are not great, the narrow MWD can be achieved only at relatively small values of $\bar{P}_w$ ($\bar{P}_z$). However, in this case the MWD distribution with boiling solvent is markedly narrower than for the polymer, obtained under the regime without boiling.

In the four-zone reactor, as in the regime without boiling, there is a possibility either to increase the maximum values of the average MW or to reach at the same time the high values of $\bar{P}_z$, $\bar{P}_w$, $\bar{P}_n$, if x_1, x_2 and x_3 are close to the values calculated according to Eqs. (54)–(56). The corresponding expressions for the average MW practically coincide with Eq. (57)–(59) (Table 13). The numerical examples for $x = 10$ and $x = 5$ are given in Table 14.

In conclusion, we shall consider the case of the continuous catalyst supply by small portions up to T_{boil}, after all the rest of the catalyst is introduced. The equation for the average MW take the form:

$$\bar{P}_w \cong \frac{x_k}{x} \bar{P}'_w + \frac{x - x_k}{x} P''_w \tag{86}$$

$$\frac{1}{\bar{P}_n} = \frac{x_k}{x} \frac{1}{\bar{P}'_n} + \frac{x - x_k}{x} \frac{1}{\bar{P}''_n} \tag{87}$$

$$\bar{P}_z \bar{P}_w x = x_k \bar{P}'_z \bar{P}'_w + (x - x_k) P''_z P''_w \tag{88}$$

where $\bar{P}'_n$, $\bar{P}'_w$, $\bar{P}'_z$ are the average degrees of polymerization of the polymer produced in all zones before boiling, and P''_n, P''_w, P''_z–in the boiling zone.

The equation for $\bar{P}'_n$, $\bar{P}'_w$ and $\bar{P}'_z$ can be found when solving the problems without boiling on the base of Eq. (72) and for the second zone:

$$\tfrac{1}{3} P''_z = \tfrac{1}{2} P''_w = P''_n = P^0_n e^{-x_k} \tag{89}$$

After the proper transformation we obtain:

$$\bar{P}_w / P^0_w \cong \frac{1}{x} \tag{90}$$

$$\frac{\bar{P}_n}{P^0_n} \cong e^{-x_k} \frac{x}{x - x_k + 1} \tag{91}$$

$$\frac{\bar{P}_z}{P^0_z} \cong \frac{1}{2} \tag{92}$$

or considering the next order of accuracy

$$\frac{\bar{P}_n}{P^0_n} \cong e^{-x_k} \frac{x}{x - x_k + 1} \left(1 + \frac{1}{x - x_k + 1} e^{-x_k}\right) \tag{93}$$

$$\frac{\bar{P}_w}{P^0_w} \cong \frac{1}{x}(1 - (x - x_k + 1) e^{-x_k}) \tag{94}$$

$$\frac{\bar{P}_z}{P^0_z} \cong \frac{1}{2}(1 + (x - x_k + 1) e^{-x_k}) \tag{95}$$

With the multi-stage method of catalyst supply, the polymer is formed with sufficiently wide distribution and absolute maximum values of all distribution moments. This absolute minimum values of all the distribution moments of the average MW and their ratios are attributed to the polymer formed by the one-time catalyst supply.

Thus, both limiting cases are the upper and lower limits of the possible variations of the MWD function moments.

To achieve a real process with characteristics close to those of the model multi-zone process, it is necessary, as was shown for the conditions without boiling, to limit the size of the catalyst portion. So, at the first stage with $0 < x \leqslant 2–3x^{1/2} e^{-x_k/2}$, catalyst portions should be small: $x_i \ll x^{1/2} e^{-x_k/2}$. At the second stage with $x < 2–3$, portions can be increased up to $x_i \ll 1$. At the third stage with $x < x_k$, catalyst portions

Table 13 The formulas for the calculation of the optimum parameters of the MW at different ways of the catalyst supply in a tubular adiabatic reactor with the boiling of the reaction mass

The number of zones	1	2	3	4
	–	$\bar{P}_{n\,max}$	$\bar{P}_{w\,max}$	$\bar{P}_{z\,max}$
	*	$x/(A \cdot (x - x_k + \ln x_k))$	*	*
$\bar{P}_n/P_n^0$	1/A	$(1 - 1/(x - x_k + \ln x_k))$	$\sim(1 + 1/x)/A$	$\sim(1 - e^{-x}k^{/2}/x)/A$
	*			
	*	*	*	*
$\bar{P}_w/P_w^0$	$1/A$	$x_k^2(1 - 2\ln x/x_k + (x - x_k)/x_k^2)/Ax$	$(1 + e^{-(x_k - 1)})/ex$	$(1 - x^{1/2}/2A^{1/2}/(xA)^{1/2}$
	*	*	*	*
$\bar{P}_z/P_z^0$	$1/A$	$x_k(1 - 3\ln x_k/x_k - (x - x_k)/x_k^3)/A$	$\sim(1 - e^{-(x_k - 1)})/e$	$1 - 2x^{1/2} \cdot e^{-x}k^{/2}$
	*			
$\bar{P}_w/P_n^0$	2	$2x_k^2(x - x_k + \ln x_k)/x^2$	$2A/ex$	$2(A/x)^{1/2}$
	*			
$\bar{P}_z/\bar{P}_w$	3/2	$3x/2x_k$	$3/2x$	$3(Ax)^{1/2}/2$
		*	*	*
x_1	x	$x_k - \ln(1 + x_k - \ln x_k)$	$1 + e^{-(x_k - 1)}$	$x^{1/2}e^{-x}k(1 - x^{1/2}e^{x}k^{/2}/2)$
x_2	–	$x - x_k + \ln(1 + x_k - \ln x_k)$	$\sim x - 1$	$\sim x - x^{1/2}e^{-x}k^{/2}$
x_3	–	–	–	–
x_4	–	–	–	–

	$\bar{P}_{w\max}\ \bar{P}_{n\max}$	$\bar{P}_{z\max}, \bar{P}_{w\max}, \bar{P}_{n\max}$	–
$\bar{P}_n/P_n^0$	$x/(A(x-x_k+\ln x_k))$	$\sim x/(A\,(x-x_k+\ln x_k))$	$xe^{-x}k/(x-x_k+1)\cdot(1+e^{-x}k/(x-x_k+1))$ *
$\bar{P}_w/P_w^0$	$\sim 1/(ex)$	$\sim 1/(ex)$	$(1-(x-x_k+1)\cdot e^{-x}k)/x$ *
$\bar{P}_z/P_z^0$	$\sim 1/e$	~ 1	$(1+(x-x_k+1)\cdot e^{-x}k)/2$
$\bar{P}_w/\bar{P}_n$	$2A(x-x_k+\ln x_k)/(ex^2)$	$2A(x-x_k+\ln x_k)/(ex^2)$	$2(x-x_k+1)e^{x}k/x^2$
$\bar{P}_z/\bar{P}_w$	$3x/2$	$3ex/2$	$3x/4$
x_1	~ 1	$\sim x^{1/2}e^{-x_k/2}$	$x_i \ll x^{1/2}\cdot e^{-x_k/2}$
x_2	$x_k-1-\ln(1+x_k-\ln x_k)$	~ 1	–
x_3	$x-x_k+\ln(x_k+1-\ln x_k)$	$x_k-1-\ln(x_k+1-\ln x_k)$	$x_i \ll 1$
x_4	–	$x-x_k+\ln(x_k+1-\ln x_k)$	–

$x=\alpha\varepsilon\Delta M/T_0^2$; $\alpha\Delta M/T_0 \ll 1$; $x_k=\varepsilon(T_k-T_0)/T_0^2$; $e^{x_k}=A \gg 1$; $x \gg 1$, $x_k \sim x$

* The second term in brackets shows the order of accuracy of the conducted estimate.

Table 14 The numerical values of the optimum MWD parameters at the different variant of the catalyst supply in a tubular adiabatic reactor with the reaction mass boiling

The number of zones	M_{max}	$\bar{P}_n/P_n^0 \cdot 10^2$	$\bar{P}_w/P_w^0 \cdot 10^2$	$\bar{P}_z/P_z^0 \cdot 10^2$	$\bar{P}_w/\bar{P}_n$	$\bar{P}_z/\bar{P}_w$	x_1	x_2	x_3	x_4
1	–	0.7	0.70	0.70	2.0	1.5	10	–	–	–
2	$\bar{P}_n$	1.4	1.75	3.50	2.5	3.0	3.2	6.8	–	–
	$\bar{P}_w$	0.7	3.50	35.0	10.0	15.0	1.0	9.0	–	–
	$\bar{P}_z$	0.7	2.40	50.0	7.0	30.0	0.3	9.8	–	–
3	$\bar{P}_w$, $\bar{P}_n$	1.4	3.50	35.0	5.0	15.0	1.0	2.4	6.6	–
4	$\bar{P}_w$, $\bar{P}_n$ $\bar{P}_z$	1.4	3.50	100.0	5.0	40.0	0.3	1.0	2.2	6.6
∞	–	1.2	10.0	50.0	~17.0	7.5	–	–	–	–

should be increased up to $x_i \ll \ln x_k$, and the obtaining x_k, all the rest of the catalyst $(x - x_k)$ should be added. The results of computations are summarized in Tables 11–14.

5.9 THE EFFICIENCY OF EXTERNAL HEAT REMOVAL IN THE REACTING TURBULENT FLOW

The intensification of heat and mass transfer in any chemical process is connected with the variation of the value D_T. Usually the increase of the linear speed of the feed flow at the fixed length of the reaction zone (1) leads to the decrease of the contact time of the polymerizate with the thermostated surface. That decrease in its turn should decrease the efficiency of external heat removal through the wall. However, the increase of V in case of fast liquid-phase polymerization leads to the essential rise of D_T, determining the effective heat and mass transfer in the turbulent flow. Therefore, the increase of D_T under conditions of external heat removal calls for a notable temperature drop in the reaction zone. Another important feature in the absence of heat removal ($\alpha = 0$) with high values of V at the initial stages of the process (1 = 1m, $\tau = 0.5$ s) is that the temperature in the reaction zone appears lower than at the lower flow speeds. Thus, the phenomenon occurs of the reaction "spreading" along the reaction zone (on axis x).

Since fast polymerization processes are characterized by inequality of the time of the chemical reaction proper τ_{ch} and of the time of transfer $-\tau_{mix}$ ($\tau_{ch} < \tau_{mix}$), it is obvious that increasing D_T decreases τ_{mix}, and both these processes become commensurable in time. The increase of the linear flow speed V, i.e., intensifying the heat and mass transfer in the system, is equal to "slowing" of the direct polymerization reaction in relation to the transfer process. Therefore traditional ways of external heat removal (low-efficient in standard technologies of fast polymerization) become notable and play a certain part when using tubular reactors with $R < R_{cr}$ both at high values of V and D_T. We can assume that external heat removal can be considerably increased due to the zone-to-zone supply of the catalyst [41, 53].

5.10 THE ZONE MODEL OF TUBULAR REACTOR UNDER CONDITIONS OF EXTERNAL HEAT REMOVAL

For the average MW and MWD of the polymer, formed in series-connected, quasi-isothermal, adiabatic (more accurately, autothermal) continuous reactors of ideal displacement and heating, the following correlations are valid:

$$\rho_w(j) = \frac{1}{\Delta M} \sum_{i=1}^{k} \rho_w^i(j) \Delta M_i \tag{96}$$

$$\bar{P}_w = \frac{1}{\Delta M} \sum_{i=1}^{k} P_w^i \Delta M \tag{97}$$

$$\bar{P}_n = \Delta M \left(\sum_{i=1}^{k} \Delta M_i / P_n^i \right)^{-1} \tag{98}$$

$$\bar{P}_z \bar{P}_w = \frac{1}{\Delta M} \sum_{i=1}^{k} P_z^i P_w^i \Delta M_i \tag{99}$$

where $\rho_w(j)$, $\rho_w^i(j)$ are the weight differential functions of the MWD of the polymer produced in the whole reactor and in the i-th zone ΔM, ΔM_i, $\bar{P}_w$, P_w^i, $\bar{P}_n$, P_n^i, $\bar{P}_z$, P_z^i are the amounts of the polymer, the weight-average, number-average, and z-average degrees of polymerization of the product obtained in the reactor and in the i-th zone respectively.

Unlike the model considered above, the temperature in each zone is determined not only by the heat balance within the reaction zone, but also by the amount of the heat removed through the wall. For the change of the temperature of the reaction mass in the cooling zone during active along the flow radius (the turbulent flow), the following equation is true:

$$\frac{dT}{dl} = \frac{2\varkappa}{RC_r \rho V}(T - \tilde{T}) = h\,(T - \tilde{T}) \tag{100}$$

Taking an integral of this equation gives the following for the temperature change in the cooling zone:

$$\Delta T(T_{k-1} - \tilde{T}) \cdot (1 - e^{-hl_k}) = \xi (T_{k-1} - \tilde{T}) \tag{101}$$

where T_{k-1} is the temperature in the preceding reaction zone or at the beginning of the considered cooling reaction zone: l_k-the length of the cooling zone; $\xi = (1 - e^{-hl_k})$ – the portion of heat removed in the cooling zone in relation to all the heat energy, stored by the system at the beginning of this zone (taking into account both the polymerization reactions and cooling in all preceding zones).

Then the temperature in the i-th zone will be equal to:

$$T_i = T_0 + \alpha \sum_{i=1}^{k} \Delta M - \sum_{i=2}^{k} \xi (T_{k-1} - \tilde{T}) \tag{102}$$

where $\alpha = q_{11}/C_r$ and the numeration of zones of cooling and reaction corresponds to fact that the cooling zone precedes the reaction zone of the same index. In its turn, the amount of the polymer formed in the i-th zone depends on the amount of the supplied catalyst A* according to the earlier kinetic scheme of isobutylene polymerization ((VI)–(X)) through the following correlation:

$$\Delta M_i = \left(M_0 - \sum_{l=1}^{k-1} \Delta M_k \right) \cdot (1 - e^{-(k_p/k_t) A_i^*}) \tag{103}$$

or if the ratio k_p/k_t does not depend on the temperature

$$\Delta M_i = M_0 e^{-k_p/k_t \sum_{i=1}^{k} A_i^*} (1 - e^{-(k_p/k_t) A_i^*}) \tag{104}$$

Knowing the temperature in each zone, we are able to find the values of the mean MW:

$$\tfrac{1}{3} P_z^i = \tfrac{1}{2} P_w^i = P_n^i = k_p^0/k_{tr}^0 \, e^{(E_{tr} - E_p)/RT_i} = k_p^0/k_{tr}^0 \, e^{\varepsilon/T_i} \tag{105}$$

where k_{tr}^0, E_{tr} are the frequency factors and activation energies of the chain transfer reaction.

Substituting (105) in formulas (97)–(99), we can find $\bar{P}_w$, $\bar{P}_n$ and $\bar{P}_z$ in Eq. (96). The function of MWD in this case is transformed as follows:

$$\rho_w(j) = \frac{1}{\Delta M} \sum_{i=1}^{k} \Delta M_i \frac{j}{(P_n^i)^2} \exp\left(\frac{j}{P_n^2} \right) \tag{106}$$

5.11 SOME CONCRETE CASES

We assume that the conditions of heat transfer are such that the most part of the stored heat in each cooling zone is removed $hl_k \geqslant 2$ (i.e., 90% or more of heat is removed). In fact, that means that the polymerization reaction proceeds under quasi-isothermal conditions at the average reaction temperature depending on the amount of the catalyst supplied in each zone: $T_i \simeq T_1 = T_0 + \alpha \Delta M$. The temperature field can be close to isothermal conditions brought in two ways: supplying the catalyst in such a way that $\Delta M_i = \Delta M$ and removing practically all heat in the cooling zone (Fig. 55, Curve 1) or, according to the definite correlation between the supplied amount of the catalyst in the i-th zone and the length of the i-th cooling zone (Fig. 55, Curve 2), by satisfying the

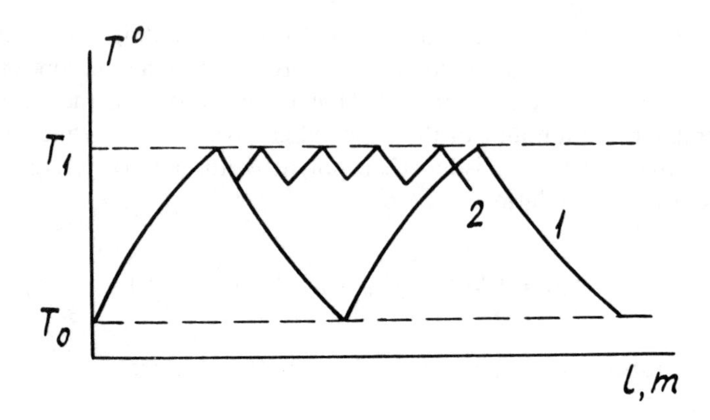

Figure 55. The diagram of the two different quasi-isothermal regime of the operation of a tubular turbulent reactor with external heat removal and multi-stage catalyst supply: $h \cdot l_k \geqslant 2$ (1) and $h \cdot l_k \leqslant 1$ (2).

condition:

$$(1-\xi_i)\cdot(T_1-T_0)+\alpha\Delta M_i=\alpha\Delta M_i \tag{107}$$

or

$$\Delta M_i/\Delta M_1=\varepsilon_i=1-e^{-hl_i} \tag{108}$$

Under such conditions, the polymer is formed with the MWD close to the exponential one and the average MW corresponds to the temperature T_1, given by the quantities of the supplied catalyst portions. In other words, the greater the average MW, the less should be the catalyst portion and the larger the cooling zone. The minimum length of all cooling zones should correspond to the regime of supplying the catalyst in small portions. In this case ($hl_k \ll 1$), the heat flow from the reaction mixture to heat transfer agent is at its maximum:

$$l_{\min}=\frac{1}{h}\left(\frac{\alpha\Delta M}{T_i-T_0}-1\right) \tag{109}$$

The quantitative evaluations of possibility of isobutylene polymerization under such conditions are given in Table 15.

The dependence of $l_{\min}$ on other parameters is obvious and expressed by formula (109). In particular, $l_{\min}$ rises with increases in flow speed and the reaction zone radius.

In the case (close to real conditions) when heat removal is not sufficient, the constant temperature in the reaction zone cannot be

Table 15 The values of the minimum cooling zone length for isobutylene polymerization with the given MW ($\rho = 0.7\,g/cm^3$, $æ = 23\,J/cm^2 \cdot s \cdot deg$; $\Delta M = 0.33$ (isobutane-isobutylene fraction); R = 2.5 cm; V = 0.5 m/s)

T_0, K	l_{min}, m	T, K	$\bar{M}_n$
180	17.0	210	22000
	5.5	250	4700
	3.0	280	2000
	2.0	300	1250
	1.3	320	620
230	24.0	250	4700
	9.0	280	2000
	6.0	300	1250
	3.7	320	820

maintained, as l_{min} is considerably higher than the real reactor design allows ($\xi_k = hl_k \ll 1$). In this case, a two-stage reactor consists of threezones: two of polymerization and one of cooling. For the temperaturein the reaction zones, we obtain $T_1 = T_0 + \alpha\Delta M_1$, and $T_2 = T_0 + \alpha\Delta M - \alpha\Delta M_1$. The equation for $\bar{P}_w$, $\bar{P}_n$ and $\bar{P}_z$ for the two-zone reactor with heat removal takes the forms:

$$\frac{\bar{P}_w}{P_w^0} = \frac{x_1}{x} e^{-x} + \frac{x - x_1}{x} e^{-x + \xi x_1} \tag{110}$$

$$\frac{P_n^0}{\bar{P}_n} = \frac{x_1}{x} e^{-x} + \frac{x - x_1}{x} e^{-x - \xi x_1} \tag{111}$$

$$\frac{\bar{P}_z \bar{P}_w}{P_z^0 P_w^0} = \frac{x_1}{x} e^{-2x_1} + \frac{x - x_1}{x} e^{-2(x - \xi x_1)} \tag{112}$$

where

$$x = \frac{\varepsilon\alpha\Delta M}{T_0^2}, \quad P_w^0 = 2 \cdot \frac{k_p^0}{k_{tr}^0} e^{\delta/T_0} = 2P_n^0 \tag{113}$$

The dependence of $\bar{P}_w$ and $\bar{P}_n$ on x_1/x–the polymer share, formed in the first zone, is extreme and is analogous to that described above (Fig. 52).

Differentiating formulae (110) and (111) to x_1 and making the derivatives equal to zero, we find the values of x_i correspond to the maximum $\bar{P}_w$ and $\bar{P}_n$. For the maximum $\bar{P}_w$ and $\bar{P}_z$ (with $x \gg 1$), only the first zone is of significance; therefore, as in the case without heat removal, we obtain $x_1 \simeq 1$ and $(\bar{P})_{\max} \simeq P_w^0/ex$. At the same time for $(\bar{P}_n)_{\max}$, the second zone, where heat removal plays a certain part, is significant, lowering the temperature and increasing the mean MW. Therefore, the equation for x_1 is changed:

$$x_1 \cong \frac{x}{1+\xi} - \frac{1}{1+\xi}\ln\frac{1+\xi+x}{1+\xi+\xi^2 x} \tag{114}$$

Correspondingly, we obtain higher mean MW values than without heat removal

$$\left(\frac{\bar{P}_n}{P_n^0}\right)_{\max} \cong e^{-x/(1+\xi f(\xi,x))}; \quad f(\xi,x) = \begin{cases} \dfrac{x}{\ln(1+x)} & \text{at } \xi \to 0 \\ 1 & \text{at } \xi \to 1 \end{cases}$$

$$\frac{P_n^0}{\bar{P}_n} \cong \left[\frac{1}{1+\xi}\left(\frac{1+\xi+\xi^2 x}{1+\xi+x}\right)^{1/(1+\xi)} + \frac{\xi}{1+\xi} + \frac{\ln\left(\dfrac{1+x+\xi}{1+\xi+\xi^2 x}\right)}{(1+\xi)x}\right] e^{-x/(1+\xi)} \tag{115}$$

The dependence of $(\bar{P}_n)_{\max}$ of the polymer on the quality removed heat ξ or the cooling zone length (1) is plotted as an example in Fig. 56.

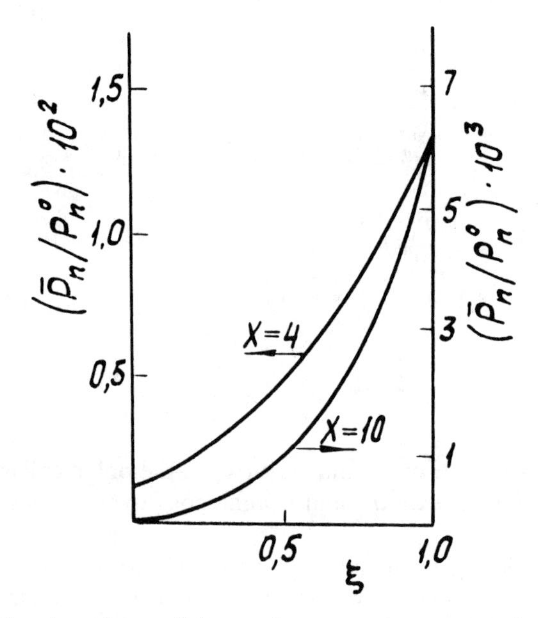

Figure 56. The dependence of the maximum number-average degree of polymerization $\bar{P}_n$ on the portion of the removed heat or the cooling zone length.

$(\bar{P}_n)_{\max}$ increases with MWD narrowing ($\bar{P}_w/\bar{P}_n \rightarrow 2$) and in the cases of effective heat transfer ($\xi \rightarrow 1$), we obtain exponential distribution. The polymer is obtained in equal portions in two zones, and the MW is characteristic of the polymer, produced at temperature $T_0 + \frac{\alpha \Delta M}{2}$, i.e. considerably higher (at $x \gg 1$) than when carrying out the reaction without hear removal. Analogous situations at three or more-stage catalyst supply are observed.

Thus, in conditions of the quasi-ideal displacement, external heat removal becomes sufficiently effective and markedly influences both the temperature field of the reaction and the MW and MWD of the polymer being produced.

6 THE DEVELOPMENT OF NEW GENERATION TECHNOLOGIES FOR FAST POLYMERIZATION REACTIONS

6.1 MODERN STANDARD INDUSTRIAL TECHNOLOGIES FOR COMMERCIAL PRODUCTION OF BUTYLENE POLYMERS

In obtaining oligo- and polyisobutylenes (polybutylenes) in the industrial production, standard practice requires raw materials (raw monomer), the content of which varies from isobutylene of high purity to isobutane-isobutylene mixtures and fractions of C_4–hydrocarbons. $AlCl_3$ and BF_3 combined with H_2O, R_2O, C_2H_5Cl, CH_3Cl etc. are used preferensy as catalysts. The process runs mainly at higher or low pressures in volumetric reactors with $V = 2 - 30\,m^3$, applying different ways of deactivation and regeneration of the catalyst and the monomer (which has not reacted, reaction heat removal, etc. The current technologies include the following general stages: 1) the preparation of the initial raw materials (cleaning, dehydration, bringing of the monomer concentration up to the required value, removal of inhibiting admixtures etc.); 2) the preparation of the catalyst (catalytic complexes or theirs combinations, the catalyst solutions etc.); 3) oligomerization or polymerization of the monomer; 4) deactivation, removal and regeneration of the catalyst; 5) degasing and utilization of the unreacted reaction components as well as of non-standard oligomer and polymer products.

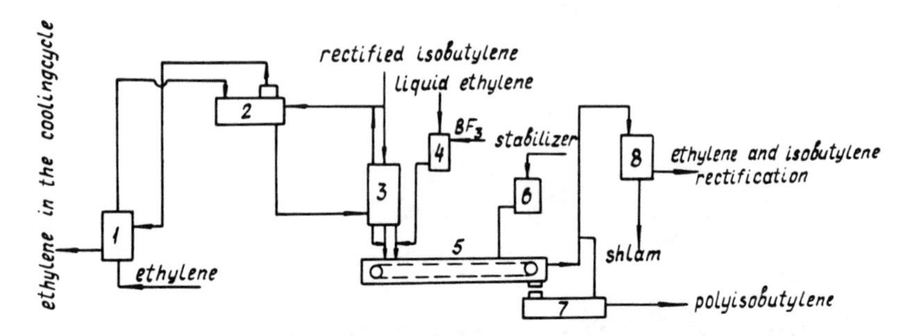

Figure 57. The diagram of isobutylene polymerization in the ethylene medium; 1,3-refrigerators; 2-separator; 4,6-measurers; 5-polymerizer; 7-degasser; 8-absorber.

Technologically, the most difficult stage proves to be isobutylene polymerization, which is strongly responsive to the slightest variation in the content and temperature of the starting raw material and catalyst, on the method of the catalyst supply in the reaction zone, etc. All the above determine the instability of the properties of the polymer formed and their frequent deviation from the technical specifications that leads to the formation of relatively great amounts of non-standard and waste products.

High-molecular isobutylene of the Oppanol type with the molecular mass of 80,000–225,000 is obtained according to the original scheme developed by "Badishe Anilin und Soda—Fabric A.G." (Germany)—on the infinite moving band with the width of 35–45 cm and the moving band length of 8–10 m (Fig. 57) in the presence of $BF_3 \cdot OR_2$ as a catalyst in boiling ethylene (the temperature is under 190 K) [2, 22, 63]. Though the process is rather simple, however, in the high-molecular polyisobutylene production, there are many unsolved problems typical of polymer production and very fast chemical processes, particularly the polymer layer growth around the infinite band that practically transforms a continuous process to a periodic one. Very important here is the ratio of the solvent to monomer (ethylene: isobutylene), which determines the rate of isobutylene polymerization temperature fields in the reaction zone and, as a consequence, the molecular mass and the molecular mass distribution (MW and MWD) of the polymer formed.

Polyisobutylene with the moločular mass of 50,000–118,000 (of Vistanex type produced by "Enjay Co., Inc.", USA) is obtained in the presence of $AlCl_3$ in the solution of ethly- or methyl chloride by polymerizing isobutylene of high purification in volumetric reactors of complex design when cooled by liquid ethylene (170–180 K) (Fig. 58) [2, 22].

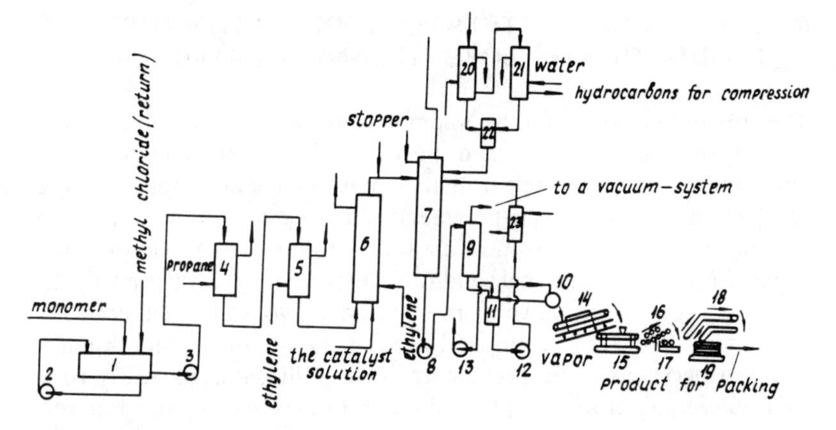

Figure 58. The technological scheme of isobutylene polymerization in the solution of methyl- or ethyl chloride; 1 – vessel for preparing charge; 2,3,8,12 – pumps; 4,5 – refrigerators; 6 – polymerizer; 7 – water degasing device; 9 – vacuum degasing device; 10 – vacuum filter; 11 – vacuum receiver; 13 – vacuum pump; 14 – drying box; 15 – injection machine; 16 – conveyor; 17 – rolls; 18 – cooling conveyor; 19 – bricketing machine; 20, 21 – condensors; 22 – separator; 23 – heater.

The kinetic peculiarities of the process in volumetric reactors are such that the temperature within the monomer-polymer particles (in microregions of growing polymer molecules) and in the microregions of the reaction zone is always higher than the reaction medium temperature. That causes the formation of the polymer with lower molecular mass and wider MWD than those resulting from the kinetic data calculations ($\bar{P}_n = V_p/(\sum V_t + \sum V_m)$].

Low-molecular polyisobutylenes and isobutylene oligomers of the "Indopol" type ("Amoco", USA) are obtained by similar procedure at 243–253 K, with ammonia used as a cooling agent. The fractions of C_3–C_4 hydrocarbons, of isobutane-isobutylene mixture (90:10–80:20), butane-butene fractions are used as a raw monomer. The solutions of $AlCl_3$ in chloralkanes, arenes, paraffins (in butane, in particular) usually serve as a catalyst. The molecular mass of the polymer formed is controlled by the temperature in the reaction zone. Due to the gradient of temperatures, the monomer concentrations and propagating polymer chains in reactors with sufficiently wide MWD ($\bar{M}_n/\bar{M}_w$—up to 5–6) are obtained and that results in the increase of the kinematic viscosity of the polymer formed.

6.2 THE NEW GENERATION TECHNOLOGIES IN THE PRODUCTION OF ISOBUTYLENE POLYMERS

The present industries of oligo- and polyisobutylene production make it possible to obtain wide range of polymers ($\bar{M}_n = 300$–$12,000$). Meanwhile, there are many problems that demand further development of the production of isobutylene polymers and others. This concerns the creation of the polymerization knot which allows us to conduct the process in strictly isothermal conditions; the recovery of waste products; the expansion of the range of raw materials bases etc. The above is determined by the fact, mentioned above (Chapt. 2 and 3), that isobutylene polymerization in the presence of electrophilic catalysts ($AlCl_3$, BF_3, etc.) proceeding at an exceptionally high rate, is accompanied in volumetric reactors by temperature fields in the reaction zone that are practically always considerably greater than the ones necessary according to the procedure (reglament).

"Torches" with various zones of high temperatures, reagent concentrations, reaction rates, etc., are formed at the points of the catalyst supply. This accounts for the complicated character of process thermostating and controlling, and, as a result, the decrease of the MW and the droadening of the MWD of the polymer formed as compared to the calculated values. In addition, the productivity of the reactors of ideal mixing decreases, the time the reaction mixture stays in the reactor becomes unreasonably long, the content of the by-products increases, etc. Therefore, model volumetric reactors of ideal mixing with a branched net of heat removing surfaces are not optimal for the fast polymerization processes, not by design, productivity, or volume.

The macrokinetic and topochemical analysis of fast polymerization (Chapt. 3) has shown that there is a complex of conditions that actively influence both the whole process and one of the main exploitating parameters of the polymer formed – its molecular mass characteristics.

Ignoring these peculiarities, one fails to obtain stable or at least rather reliable reproduction of polymers with the desired properties.

The basic factors that make it possible to secure the control over very fast chemical processes of liquid-phase polymerization are depicted by scheme [64].

However, we do not obtain a full understanding from only the consideration of the dependence of the polymer MW and MWD on the temperature raw monomer and the reagent concentration parameters in the classical analysis of kinetic laws according to the Mayo–Lewis equation. For this reason, that dependence cannot serve as the basis for predicting the properties of the obtained product, many other important

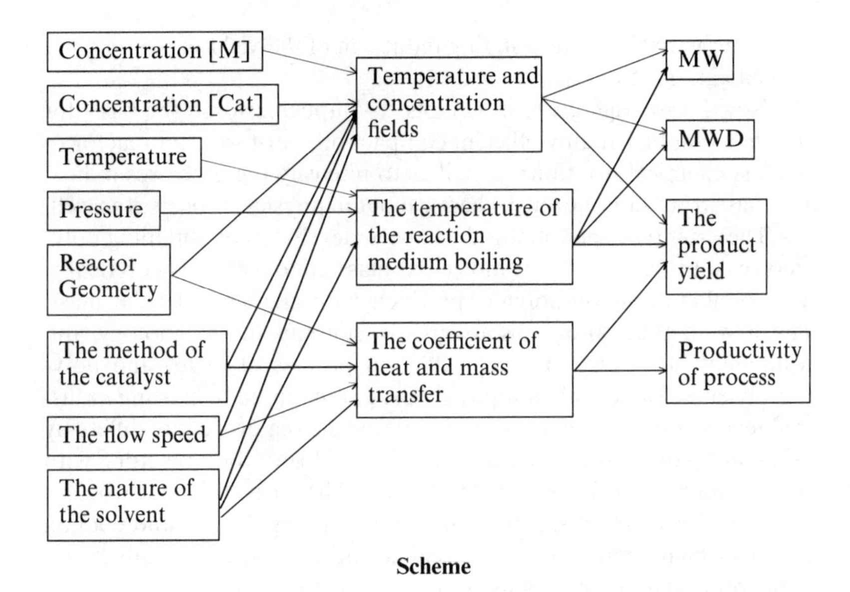

Scheme

factors also determine the numerical values of the MW and MWD. As a consequence, in studying the dependence of the MW on the feed temperaure of raw monomers, one can not obtain the true result only due to the fact that the coefficient of turbulent diffusion D_T will be different in each particular case. Meanwhile, D_T is connected with the flow speed, the method of reagent supply, etc. It determines the optimum design of the basic reactors (Sections 4.3 and 4.4, Chapt. 4).

It should be particularly emphasized that there exists a distinct dependence of the numerical values of the MW and MWD of the products formed, as well as of the process selectivity, on the linear rate movement of the feed flow (V). This has never been taken into consideration before, although to stabilize the reagent supply in the flow is far simpler than to achieve uniform and practically instantaneous catalyst distribution in the reaction mixture volume in fast chemical processes, even at very intense mixing. At the same time, the speed of the raw monomer flow affects the temperature field of the reaction zone as well. Incidentally, external heat removal (external thermostating) in realizing fast chemical processes in modern production appears inefficient. When it is necessary to obtain the product with the normal MWD in one apparatus, limitation of the reaction zone is needed because, mentioned above, the reaction proceeds in a form of a "torch." Beyond the "torch" in the current volumetric reactors there is usually a zone where the process does not proceed at all, and there is a non-attendance slip of the monomer with the corresponding reduction of the yield of the polymer formed [31–34, 54, 65].

New knowledge given in Chapts. 1–5 appears to be the basis for creating a unified, highly-efficient compact tubular turbulent reactor to run fast chemical reactions as well as to intensify the processes of heat and mass transfer not only in chemical but in physical processes as well.

The essentially new method has been developed of obtaining polymers of isobutylene with the molecular mass up to 50000. Theoretically, it is possible to obtain polymer products with greater molecular mass. The new method allows us to greatly simplify the technology and equipment furnishing of the production process of oligo- and polyisobutylenes. Sufficiently unique, simple as compared to existing industrial technologies, the energy- and material-saving procedure differs by using not volumetric reactors with $V = 2–30\,m^3$, but reactors with continuous polymerization in ideal mixing. These new tubular turbulent reactors of very small size without special mixing devices and cooling systems which operate in intense regime and provide high productivity of the process and high quality of the polymer formed.

The base characteristics of a new reactor compared with the ones in current use are given in Table 16 [2, 64, 66, 67, 72].

The new technology characterized by usage of small-size tubular turbulent reactors provides the reduction of raw materials spent as well as the decrease of the specific weight of side reactions and secondary processes (the consequence of the exceptionally short time the reacting mass stays in the reaction zone). The process on the whole is distinguished by higher environmental harmlessness and greater safety.

The use of a tubular turbulent reactor makes it possible to utilize a basically new process procedure and to simplify the equipment by making it universal and the process, including heat exchange easily controllable. The new method of very fast polymerization allows us when necessary, to stabilize the reaction temperature conditions and to conduct the process under quasi-isothermal regime, irrespective of intensity of heat evolution and the chemical reaction rate. The apparatus is simple to manufacture and needs no special mixing and cooling devices. It operates on the principle of stabilizing the turbulent vortex with using tubular reactors. The new method keeps to the main principle of working technological procedure, reduces the production areas (the required area for the reactors is small), increases total productivity at least five-fold (the upper limit is not determined), increases specific reactor productivity more than 1000 times, raises the possibility of using raw monomer of wide ranging composition (from 5% up to 80% and more of concentrates of isobutylene) as well as of using non-standard

Table 16.

Reactor parameter	Tubular turbulent reactor (BSU, Ufa; ICP, RAS Moscow).	Volumetric reactor (LenNII, Khimmash, Russia	Volumetric reactor of "Stratco", (USA)	Volumetric reactor of "Amoco", (USA)
Volume, m^3	$0.02 \div 0.06$	$1.5 \div 4.0$	up to 29	20–30
Metal capacity, (t)	0.05	7.5	up to 40	50
Energy capacity, (rel. units)	$0.08 \div 0.85$	1.0	1.0	1.0
Time for the regime exist, (hr)	0.01	$5 \div 6$	$3 \div 5$	3–5
Time of the reaction mass staying, (hr)	0.003	$1.5 \div 2.0$	1	1
Productivity, (t/hr)	more 8–10	$1.5 \div 2.0$	up to 5	up to 10
Specific productivity, (rel. units)	more 1000	1	1	1–2
Specific expenses, of (rel. units):	0.8	1	1	1
feed	$0.7 \div 0.8$	1	0.9	1
catalyst	$0.5 \div 0.6$	1	1	1

raw monomer without additional treatment (the result of a markedly lower sensitivity to harmful admixtures), reduces the number of operating personnel, the duration of a number of auxiliary operations and the time needed for the overhaul, decreases if necessary, of the width of the MWD in the polymer formed (from 8–12 down to 2.5–3), adds versatility to the process (in particular, obtaining in a single unit a wide range of polymers with $\bar{M}_n$ from 200 up to 50,000 and more).

Certain data on the industrial production of isobutylene using small tubular turbulent reactors are presented in [64, 66, 68–75].

6.3 THE INDUSTRIAL EXPERIMENT WITH POLYBUTENE IN A SMALL-SIZE TUBULAR TURBULENT REACTOR

A tubular turbulent reactor has been widely tested and introduced in the production of isobutylene polymers (oligo- and polyisobutylenes, low-molecular butyl rubber) in Russia, Ukraine and Azerbaijan.

Table 17. The fractions of C_4 -hydrocarbons of butylene polymers in a turbulent reactor

Composition of fractions of C_4 -hydrocarbons	Hydrocarbon fraction, mass%		
	isobutane-isobutylene (IIF)	butane-butylene (BBF)	butene-isobutylene (BIF)
n-butane	1–2	10–31	—
isobutane	68–50	42–59	—
isobutylene	50–30	5–11	40–45
α-butylene	—	2.8–10	27–30
trans-butylene-2	—	2–7	8–10
cis-butylene-2	—	4–10	3.5–4.5
C_3—C_5 hydrocarborns	2–1	2–9.5	5–15
butadiene-1, 3	—	—	0.1–0.2

The conducted research and industrial experience have confirmed all the advantages of a tubular turbulent reactor that were described in Section 6.2. The tubular turbulent reactor was compared with the standard volumetric reactor widely used at present in the production of isobutylene polymers, with the polymer products that meet the standards and technical demands being possible to obtain using different raw materials (for example-fractions of isobutane-isobutylene or butane-butylene) (Table 17) and catalysts (the solutions of $AlCl_3$ in ethyl chloride, chlormethylene, butane, alkyl aromatics etc.)

When the isobutane-isobutylene fractions are used, the dependences of isobutylene conversion on the $AlCl_3$ content in ethyl chloride at various temperatures of the raw material at the tubular turbulent reactor inlet ($d = 0.05$ m, $l = 2$ m, with productivity of 3.8 m^3/hr) are given in Fig. 59. The temperature and the reagent concentration regions have been chosen so as to obtain oligoisobutylene with $\bar{M}_n = 700–1200$. It is apparent that the raw material temperature rise at the reactor inlet increases the influence of the catalyst amount supplied in the reacting mixture.

The temperature increase in the mixing zone causes the production of polymers with lower molecular masses (the number of the macromolecules formed grows) (Fig. 60).

It should be noted that the outlined dependences are characteristic of the final condition of the system. The isobutylene polymerization process is obviously more complicated. The MWD width of the polymer product formed is apparently dependent on the size $\Delta T = T_p - T_0$, i.e.,

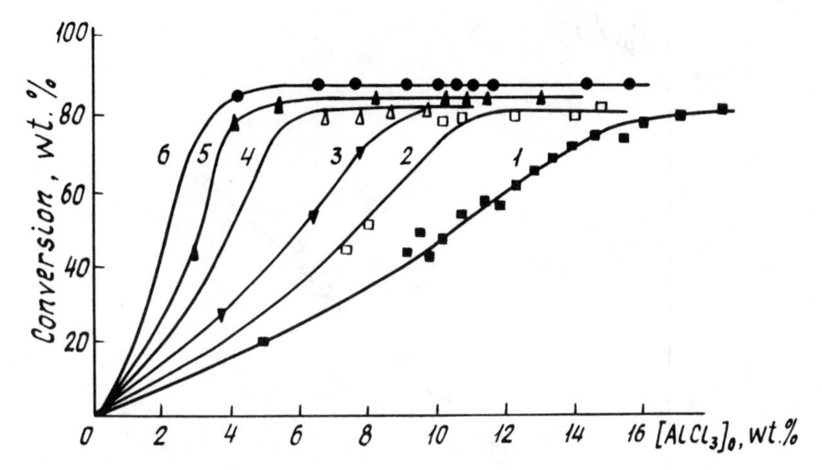

Figure 59. The dependence of the isobutylene conversion on the catalyst concentration ($AlCl_3$) at the feed temperature at the reactor inlet T_0, K: 1 – 278; 2 – 273; 3 – 263; 4 – 253; 5 – 248; 6 – 238.

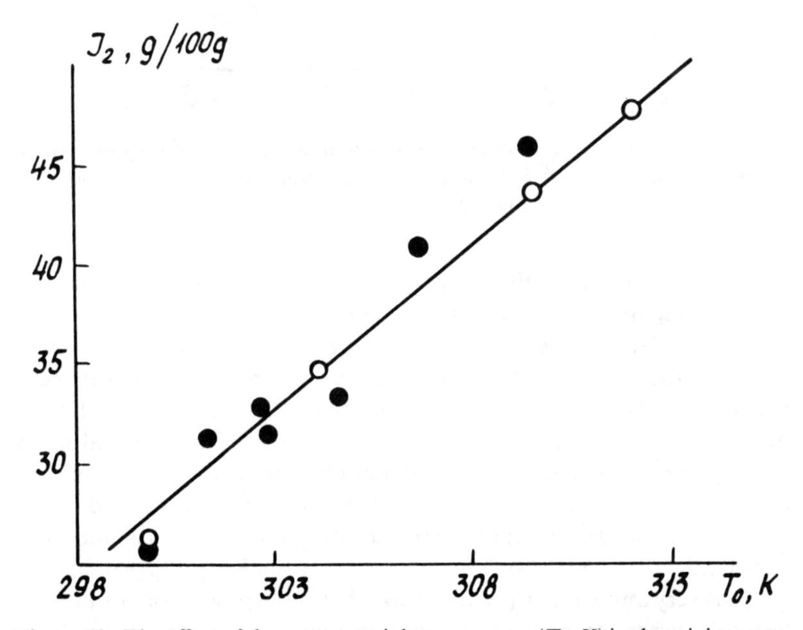

Figure 60. The effect of the row material temperature (T_0, K) in the mixing zone on the iodine number (I_2) of polybutene.

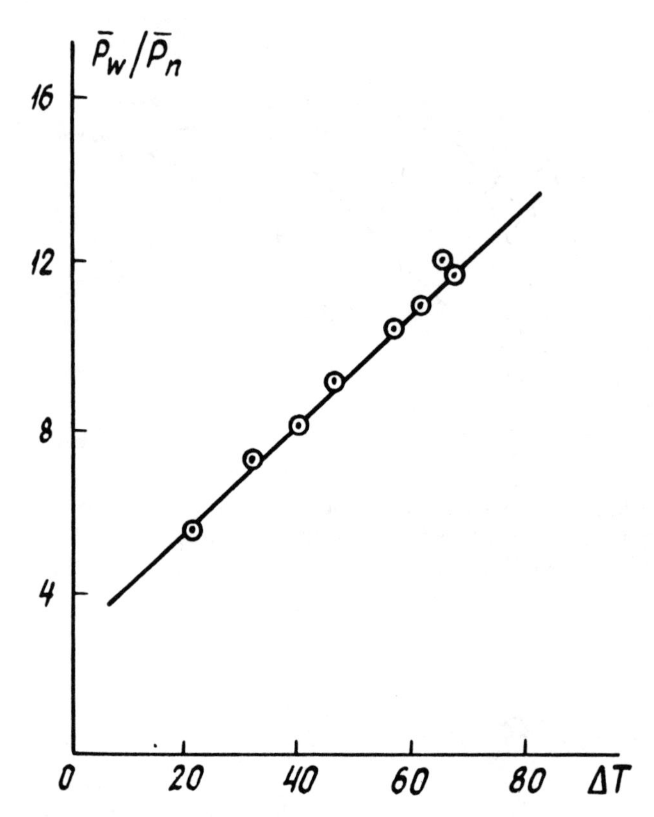

Figure 61. The dependence of polydispersion index ($\bar{P}_W/\bar{P}_n$) of polymer on the temperature difference (ΔT) between the temperature in the reaction zone (T_p) and the one at the reactor inlet (T_0).

the difference between the temperatures in the reaction zone and the initial temperatures of raw material (Fig. 61). Within a wide range, this lets us regulate the MWD width and accounts for one of the main reasons (besides the existence of the fields of temperature and reagent concentrations along the reactor coordinates) of the formation of polymer products with abnormally high values of the MWD width. In isobutylene polymerization, the reaction zone temperature at the inlet in a standard volumetric reactor of ideal mixing in the current industries changes by 50–100° in spite of the seemingly intense mass and heat transfer.

It is very important to bear in mind that, if the process of isobutylene fast polymerization proceeds under the conditions far from the macrokinetic regime of quasi-ideal displacement in turbulent flows (type

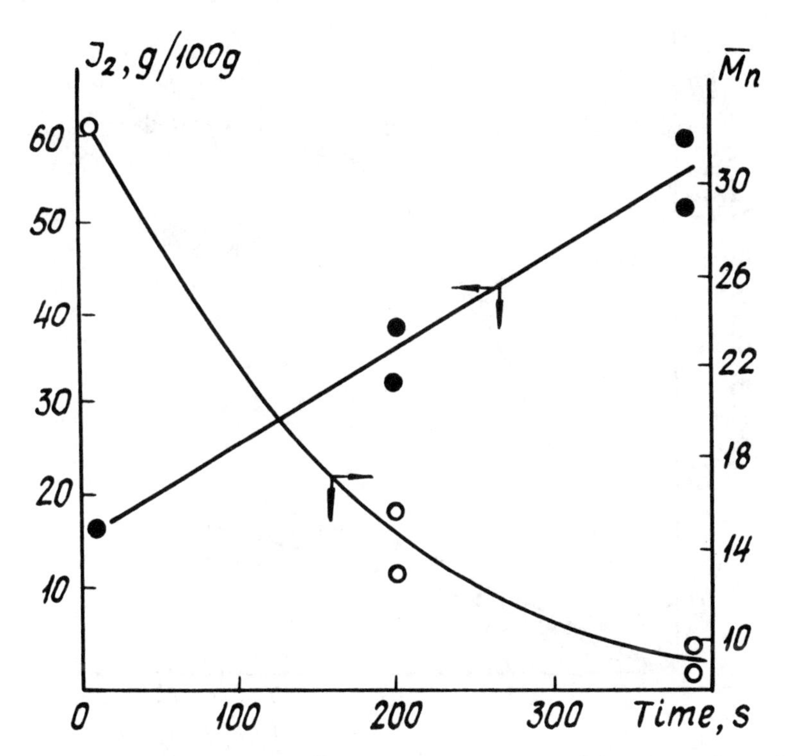

Figure 62. The dependence of $\bar{M}_n$ and iodine number (I_r) of the polymer formed on the reaction time (τ, s).

A regime, Section 2.1, Chapt. 2), then the additional polymerization of the rest of the monomer at continuously growing temperatures with moving away from the point of the catalyst and monomer supply in the tubular reactor leads to the increase of unsaturation of the polymer formed and, accordingly, the drop of its $\bar{M}_n$ (Fig. 62). The decrease of the MW and the broadening of the MWD of polyisobutylene occur mainly due to the accumulation of the low-molecular fraction. In particular, up to 2–3% of oligomeric isobutylenes ($\bar{M}_n = 200$–400) are formed with the separation from the final product of the reaction being conducted in a vacuum column.

The butane-butylene fractions (BBF) are appropriate to use for the synthesis of polybutylenes, composed of a mixture of polyisobutylenes and polybutene-1. The main difficulty in the process is that butylenes isomere differ considerably from each other in their reactivity (Fig. 63), that does not allow us to involve them completely in the polymerization process using standard volumetric apparatus ($V = 2$–$30\,m^3$) equipped

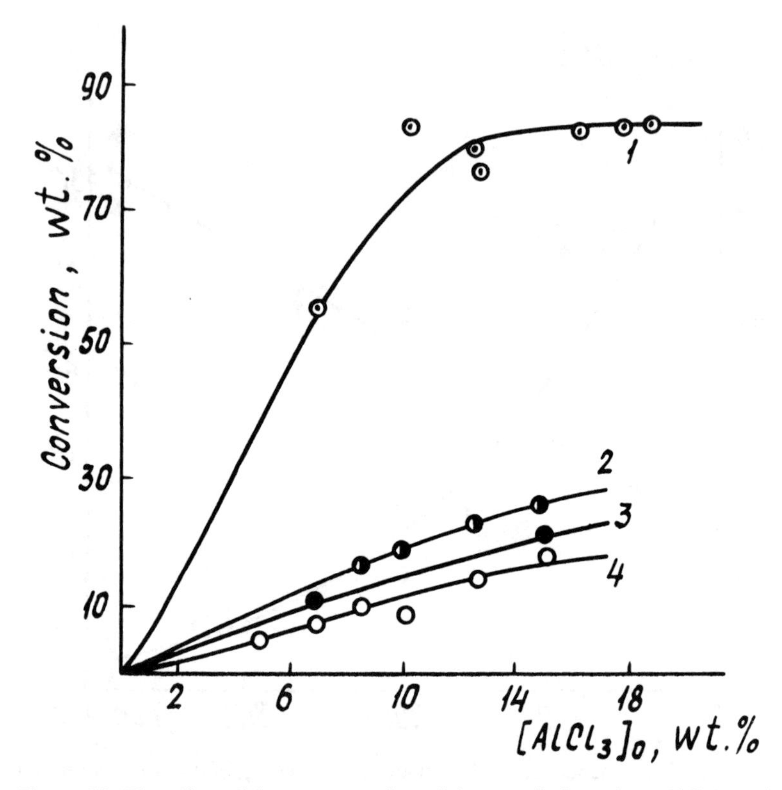

Figure 63. The effect of the concentration of the supplied catalyst ($AlCl_3$) on the conversion of olefin components of butane-isobutane fraction: 1 – isobutylene; 2 – butene-1; 3 – trans-butene-2; 4 – cis-butene-2.

with the intense stirring and thermostatic devices to remove the reaction heat. The volumetric reactor design with the polymerizate circulating and the constant supply of raw monomer creates conditions for the predominant polymerization of the most active monomer-isobutylene.

Actually, one may obtain in practice the selective polymerization of isobutylene. In the current industries this fact made it possible to carry out two-stage polymerization schemes in volumetric reactors of ideal mixing. Isobutylene and a relatively small amount of n-butylenes react at the first stage. At the second stage (at higher temperatures) the rest of the isobutylene and the most of butylenes react.

Since the butylene polymerization in volumetric reactors is characterized by temperature and reagent concentration differences, the quality of the polymer formed decreases due to the formation of greater amounts of the low-molecular fraction and polymer products with the broader MWD. In addition, the drivers of the stirring devices are frequently out of order, and hard-to-reach inner equipment of the reactor is subject to intense corroding.

But tubular reactors work differently. In an intense turbulent regime the reaction of polymerization ($Re > 10^4$) provides the process run in the absence of volumetric temperature and concentration gradients of the reagents, i.e., under the conditions close to the isothermal ones. This makes it possible to obtain a series of polybutenes that meet technical demands, in particular, on the component for producing lubricating and cooling liquids, electroinsulating liquids and the dispersion medium component.

The basic technical and economical parameters of a working tubular reactor are far superior to those involved with standard commercial volumetric reactors of ideal mixing. Productivity by feed is 2–2.5 times greater at a considerably smaller volume of the reaction space (75–100 times) and model capacity (150 times). The specific expense of the catalyst in a tubular reactor ($AlCl_3$) is reduced in 1.5–1.6 times. The butylenes conversion degree reaches 95–100% at one stage. In addition, due to the absence of stirring devices, the electrical power consumption drops by 20–25%.

A tubular turbulent reactor is characterized by the reduction of the light polymer yield, which is the production waste. The polybutenes formed are characterized by narrower molecular mass distribution than the polybutenes of the same grade obtained by the conventional procedure. As a result, the expense of raw monomer (BBF) decreases by 14–16%. All these savings, plus reduction of maintenance costs, decrease the first product cost of polybutenes. It is still more difficult to use a cheaper butylene-isobutylene fraction, which found no appropriate application in the process of obtaining oligomers and polymers of butylenes as the raw material for producing useful isobutylene polymers according to the standard procedure in volumetric reactors, due to the inhibiting effect of butylenes and divinyl [71, 74]. The use of a tubular turbulent reactor makes it possible at one stage to obtain not only low-molecular butylene polymers (Octoles-K with $\bar{M}_n = 700–1500$) [74] but also higher-molecular polymer products as well—thickeners and additives of the KP-5 grade with $\bar{M}_n = 4000–6000$ and those of the MP-10 grade with $\bar{M}_n = 9000–15000$ that are, in fact, the solutions of polyisobutylene in the base oil.

The low raw monomer temperature at the inlet into the reaction zone (173 K) in tubular turbulent reactors provides nearly complete isobutylene polymerization proceeding in the zone of mixing the monomer and catalyst. In this case the polymerizate containing the butylene mixture is heated due to the reaction heat, up to 233–243 K. This fact, combined with the absence of the raw material recirculation, provides the conditions for reacting for all butylene isomers (n-butylene-1, cis-and-trans-butylenes-2). Butylene conversion in a volumetric reactor during a one-step exceeds 25–35%, while in a tubular turbulent reactor it approaches 85–95% by weight.

The low rates of polymerization of n-butene as well as of trans-and-cis-butenes-2 (Fig. 63) make it necessary to use as a catalyst of a large amount of $AlCl_3$ (or another suitable electrophylic catalyst).

The dependences of isobutylene conversion on the $AlCl_3$ concentration at various raw monomer temperatures at the reactor inlet are given in Fig. 64. The given temperature and expense of catalyst values are chosen on the basis of the necessity of obtaining Octole-K which meets the technical demands. As the temperature of raw monomer at the reactor entrance increases, the catalyst expense effect becomes more evident.

At the entrance raw monomer temperature of 268 K, the isobutylene conversion increases twice, from 45 to 90 per cent by weight with the growth of the aluminum chloride amount from 0.09 up to 0.16 per cent by weight on the raw material. And by contrast, within the temperature range at the entrance in reactor of 238–248 K and the catalyst consumption of more than 0.04 per cent by weight, the isobutylene conversion

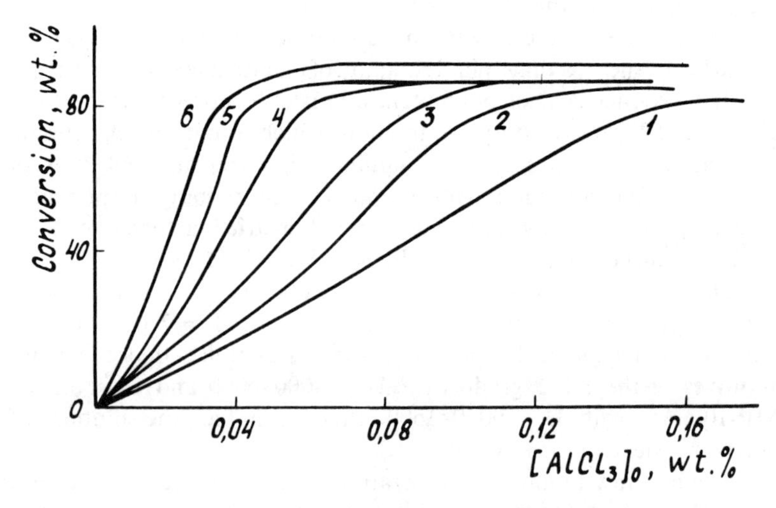

Figure 64. The dependence of the isobutylene conversion on the catalyst consumption ($AlCl_3$) at different temperatures of the raw monomer at the reactor inlet, T_0, K: 1 – 278, 2 – 273, 3 – 263, 4 – 253, 5 – 248, 6 – 238.

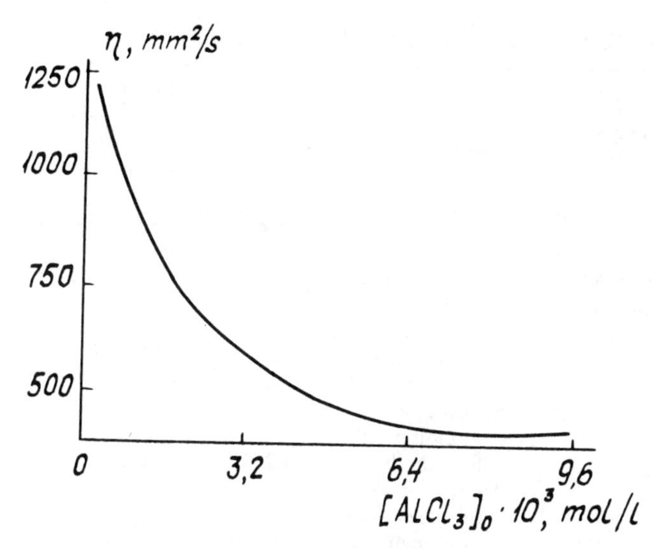

Figure 65. The effect of the catalyst dischange ($AlCl_3$) on the viscosity (η) of polybutene (373K) at the raw monomer temperature at the reactor inlet, $T_0 = 248$ K.

dependence curves turn to straight lines with negligibly low tangent of the bend angle. In the indicated area, the conversion exceeds 90 per cent by weight and is practically independent of either catalyst expense or temperature.

The minimum catalyst expense is determined by an upper border of technical requirements to Octole-K in viscosity, i.e., 0.06 m^2/s. Minimum catalyst expense is 0.04 per cent by weight on raw monomer (Fig. 65). With more catalyst consumption, this parameter effect is rather weak, but the standard product can be obtained within the wide range of its values. The upper value is determined mainly by material saving. As the process temperature grows, the polybutenes yield increases but its molecular mass drops.

The regimes for obtaining polybutenes in tubular turbulent reactors in comparison with volumetric ideal mixing reactors are given in Table 18.

Of special interest is the influence of the reaction polymerization time on the butylene conversion depth (Table 19).

If, in volumetric reactors, the time of the polymerizate staying in the first process stags is from 7000 to 8000 s, and the conversion by the sum of butylenes does not exceed 35 per cent by weight, in a tubular turbulent reactor the whole bulk of isobutylene and up to 30 per cent by weight of n-butylenes are polymerized already in the zone of mixing of the raw

Table 18. Technological regimes of polybutene production

Polybutene grade, GOST, TU (Russia)	Parameters	Reactor: a full volumetric mixing, 1 stage	Reactor: a full volumetric mixing, 2 stage	Reactor: tubular
Electroinsulating liquid, Octole* (GOST 12869–77)	Temperature, K			
	at the reactor entrance	—	—	270–268
	in the reactor	248–245	248–255	273–283
	Pressure, MPa	0.45–0.5	0.45–0.5	0.45–0.5
	Expense of raw monomer in flow m^3/hr	1.2–1.5	1.2–1.5	1.2–1.5
	$AlCl_3$ expense on the raw monomer, % wt	1.3–1.4	1.5–1.6	0.85–1.06
	The exposure time, sec	7000–8000	7000–8000	50–100
	Conversion, % wt			
	of isobutylene	85–95	100	100
	$\sum$ butylenes	25–35	85–98	85–98
Polybutene for LSL (TU 38.101713–81)	Temperature, K			
	at the reactor entrance	—	—	265–263
	in the reactor	286–291	291–296	318–323
	Pressure, MPa	0.4–0.45	0.4–0.45	0.4–0.45
	Expense of raw monomer in flow m^3/hr	1.2–1.5	1.2–1.5	1.2–1.5
	$AlCl_3$ expense on the raw monomer %wt	0.7–0.95	0.9–1.15	0.9–1.2
	The exposure time, sec	7000–8000	7000–8000	50–100
	Conversion, %wt			
	of isobutylene	90–100	100	100
	$\sum$ butylenes	31–72	83–96.5	85–97

* Polybutene TU 38.101679–74 is produced in the same conditions

Table 19. Butylene conversion in the mixing zone (the first point of sampling) and at the 20 m distance from the mixing zone (the second point of sampling)

Parameters	Experiments		
	1	2	3
Hydrocarbon content, per cent by weight:			
C_3H_6	0.3	0.2	2.0
C_3H_8	6.0	1.9	2.4
iso-C_4H_8	1.5	3.3	7.6
α-C_4H_8	3.2	7.8	9.9
cis-C_4H_8	3.2	8.5	9.5
trans-C_4H_8	1.7	5.4	7.9
iso-C_4H_{10}	73.7	59.8	43.9
n-C_4H_{10}	10.2	12.3	16.8
C_5-hydrocarbone	0.2	0.8	0
Conversion, per cent by weight:			
iso-C_4H_8	100/100	83/100	97.2/100
α-C_4H_8	51.2/97	24.3/34.6	27.7/80.7
cis-C_4H_8	21.9/68.3	38.2/72.7	17.8/66.8
trans-C_4H_8	40.6/78.8	37.0/83.7	19.5/83.8
Contact time, sec	14.0/94.3	15.0/105.0	15.0/101.0

* Numerator—the first point, denominator—the second point of polybutene sampling.

monomer and catalyst flows. The reaction time is 8–15 seconds, and in 60–100 seconds the reaction is almost completed. The depth of the involvement of n-butylenes in the polymerization process exceeds 90 per cent by weight.

On the average, to obtain, for example, polybutenes of the KP-10 grade from the butadiene-isobutylene fraction ($\bar{M}_n = 9000$–15000), the process should be run at temperatures below 243 K and the catalyst concentration (the $AlCl_3$ solution in alkyl aromatics) should be less than 0.03 per cent by weight. To obtain polymers of the KP-5 grade ($\bar{M}_n = 4000$–6000), it is necessary to conduct the process at 253–263 K with a catalyst concentration of 0.1–0.2 per cent by weight. All the advantages of a turbulent tubular reactor over volumetric reactors of ideal mixing are retained at the same time.

6.4 SOME CONSEQUENCES

A tubular turbulent reactor appears to be universal. It has become the basis for developing many new technologies, not only in realizing very fast polymerization processes but also in running fast chemical processes with low-molecular compounds and in physical processes where the mass exchange proves to be the leading stage too.

A series of tubular turbulent devices have been developed and introduced in the production of synthetic rubbers, including the stereoregular ones. When obtaining synthetic divinylstyrene rubbers of the grades SRMS-15, SRMS-30 and SRMC-50 (emulsion copolymerization of butadiene with α-methylstyrene) tubular turbulent apparatus are used at the stages of preparing emulsions, mixing the emulsions with the initiator, mixing latex with oil (SRMS-30, SRMS-15), stoppering the process.

The efficient turbulization of the reaction mass flow in a tubular turbulent apparatus leads to the formation of a uniform mixture in all reactor batteries and, as a consequences, to the uniformity of the product obtained. The new technology, together with the application of small-size tubular turbulent apparatus, improves the technological parameters of the process and makes it sufficiently controllable. It ensures the marked increase of the technological cycle duration (till the shut-down of the equipment for cleaning the reactors-polymerizers)—twice or more times, the increase of the final product yield, the improvement of some technical parameters of the rubber, and the uniformity of the polymer within a single batch.

In the production of stereoregular 1,4-cis-polyisoprene synthetic rubbers of the ISR-3 grades (the catalyst system is based on organoaluminum compounds and titanium chloride) and ISR-5 (the catalytic system is based on organoaluminum compounds and chlorides of rare-earth elements), tubular turbulent reactors are used at the stages of preparing the catalysts, mixing the catalyst with the monomer, introducing the stopper into the reaction mixture to complete the reaction, and decomposing isoprene hydrochloride and the catalyst. The new method of using tubular turbulent apparatus cuts catalyst expense by a factor of 2 to 4 or more (ISR-5); increases obtaining of polymer product with the polydispersion index (MWD) up to 3 ± 1; sharply reduces the light fraction content in the final product (ISR-3), the strict correspondence (for the first time) of the rigidness to the plasticity (ISR-5). It makes it possible to obtain rubbers of various grades at the same device.

The obtained products (especially ISR-5) are close to natural rubbers in the combination of their characteristics property. They are in full agreement with the theoretical assumptions (Table 20).

The use of small tubular turbulent reactors operating under the conditions of quasi-ideal displacement of 1, 2-dichloroethane and ethyl chloride made it possible to carry out basically new methods of liquid-

Table 20. Qualitative parameters of stereoregular polyisoprene rubbers

Parameters	Basic process	New process using tubular turbulent reactors
The catalyst expenses, rel.units		
ISR-3	1	0.5
ISR-5	1	0.3
MWD	5–7	3–4.5
Stereoregularity, %	91 ± 2	97 ± 2
The product uniformity, i.e., the fraction content with the deviation by MW and MWD, per cent by weight	8–15	2–3
The crusting rate, rel. units	1	0.1
The oligomer content in the product, per cent by weight	13 ± 2	2 ± 2

phase catalyst chlorination and hydrochlorination of ethylene that notably simplify the technology and equipment necessary for the process.

The size of a tubular reactor as compared to the commercially available volumetric reactors is reduced by a factor of 5 to 10 and more times, since it has neither heat exchange equipment for external cooling of tubular reactors nor stirring devices. The process proceeds in intense regime with high selectivity with the spent chlorine being used as a chlorating agent, which markedly reduces the environmental dangers of chlorine production.

The new technology in the production of ethyl chloride, dichloroethane has confirmed all the conclusions about using a small tubular reactor, the conservation of the basic principles of the current technology of dichloroethane and ethyl chloride production, the reduction of the industrial premises, the decrease of the metal capacity of a reactor by 20 to 100 times, the increase of the reactor specific productivity by not less than 10 to 100 times, and the drop of the reaction mass staying time by not less than 60 to 120 times, which leads to a marked decrease in the amount of chlorine-containing side-products formed.

The production of methyl-ethyl ketone has been modernized using tubular turbulent reactors. In today's industries, when butyl sulfuric acid is prepared, volumetric reactors of mixing or hermetic pumps for chemically aggresive media are used. Tubular turbulent reactors need no such devices, making the technology simpler and simultaneously increasing the degree of sulfuric acid saturation with butylenes, decreasing the expenses rates of the butane-butylene fraction and sulfuric acid per a ton of the finished product, reducing polymerization in steam-stripping columns, and expanding the equipment's maintenance-free period. This leads to a considerable power saving due to the removal of the power-consuming equipment from the mixing block.

Tubular turbulent reactors proved to be efficient in realizing physical processes of various types: mixing, dispersion, emulsification, extraction.

The tubular turbulent apparatus is convenient for mixing two liquids, especially those differing in density and viscosity (water-glycerol, water-sulfuric acid etc.). In addition, the dispersion of two liquids that differ in density and viscosity presents certain often unsurpassable difficulties, which increase as do the differences in the liquid densities and viscosities.

High-dispersion uniform (homogeneous) emulsions (10–15 per cent by weight) with the drop size of 0.5 ± 0.3 mm are produced in tubular turbulent reactors at the density differences from 0.5 to 2 g/cm^3 and/or the viscosity differences—from 2 to 300 cp at the main flow speed of 0.5–1 m/s.

Unlike volumetric reactors with stirring devices, tubular turbulent reactors have no stagnation zones; the dispersion phase is uniform throughout the mixing zone. The possibility of stabilizing the turbulent vortex along the reactor length (without the reactor length limination) makes it possible to stabilize the emulsion.

Using tubular turbulent reactors of the original design at one chemical plant, the problem of extracting phenols and pyridines from the resin has been solved.

7 CONCLUSIONS

Fast polymerization processes proceeding at very high rates ($k_p > 10^3$ l/mol·s) with a characteristic reaction time (τ_{ch}), faster than the rate of the reactant transfer ($\tau_{ch} > \tau_{mix}$) require specific approaches both to fundamental and applied research and to technological design. Fast polymerization processes are characterized by basically new phenomena. The example of isobutylene polymerization has shown that the reaction proceeds under extreme conditions. The molecular mass characteristics of the polymer formed depend on the catalyst and monomer concentration, though the kinetic scheme of the process predicts the independence of the MW and the MWD on the concentration of both the catalyst and monomer.

Under standard conditions of isobutylene polymerization, especially in volumetric reactors, the process is characterized by the formation of fields of temperature as well as of the monomer concentration and propagating polymer chains along the reaction żone coordinates. As a consequence, the MW decreases and MWD broadens in comparison with the most probable figure, which is characteristic of the isothermal process conditions. The external heat removal does not prove to be sufficiently effective in general.

The obtained results lead to the conclusion that fast liquid phase processes should be considered as an independent class of chemical reactions.

The most similar reaction class appears to be the processes of burning, which are distinguished as a separate class of oxidation reactions due to the characteristic macrokinetic peculiarities.

It is possible to refer to fast polymerization processes to a special class of chemical reactions thanks to the presence of the following distinguishing features [54, 76, 77]: 1) the objects of investigation, in particular, very fast polymerization processes, are characterized by very fast chemical reaction compared to that of transfer; 2) the specificity of investigation of fast chemical processes in the turbulent flows limited by inpenetrable wall; 3) the specific nature of the chemical process control, including the molecular-mass characteristics of the products formed (the flow rates, turbulization parameters, mixing characteristics, the efficiency of mass and heat transfer, the reaction zone geometry etc.); 4) new fundamental laws, in particular: a) the existence of the new and analogue-free macroscopic regime of quasi-ideal displacement in turbulent flows that makes it possible to conduct processes under quasi-isothermal regimes without heat removal; b) the evident connection between the critical radius of the reaction zone (R_{cr}), determining the upper limit of the macroscopic regime of quasi-ideal displacement in turbulent vortices, and the turbulent diffusion coefficient (D_T) and the rate constant of the active centers termination (k_t); c) new possibility for the evaluation of effective constants of chain propagation and termination k_p and k_t, in particular, when the reaction zone radius $R < R_{cr}$ and the time of the reaction mass staying in the reaction zone $1/V$ change; d) the possibility of controlling the parameters of mixing and heat transfer by varying the reaction zone geometry (R and 1), the turbulent diffusion coefficient (D_t), etc.; 5) as a consequence, a new class of chemical reactions—fast polymerization processes—in practice should be carried out according to new specific technology, in particular, using tubular reactors in turbulent flows.

The dependence of the molecular-mass characteristics of the polymer products on the method of the reagent supply into the reaction zone and efficiency of the flow mixing, especially in mixing the liquids different in density and viscosity, proves interesting from the scientific and, most important, from the practical points of view.

The optimum process of mixing requires the active particle transfer, commensurable with the rate of the chemical reaction itself ($\tau_{mix} \cong \tau_{ch}$).

With the other conditions being equal—the radial way of the catalyst supply into the reaction zone and the cocurrent way of the monomer supply as compared to the cocurrent way of supplying both the catalyst and the monomer, the flow turbulization sharply increases, markedly improving the efficiency of mixing the liquid flows as well.

There is an increase of D_T, the reaction zone is "compressed", the monomer conversion increases and the molecular-mass characteristics of the polymers formed are improved.

Another convenient and effective factor that makes it possible to control very fast polymerization processes in turbulent liquid flows is the change in the method of introducing the catalyst supply into the reaction zone—the fractional supply. This is achieved by using either the zone model of a tubular reactor or the cascade of tubular reactors with the catalyst supply into each of them.

The control of the variation of the turbulent diffusion coefficient D_T in the reaction zone is noteworthy. If D_T is not kept constant in the flow direction (irrespective of the method of the reagent supply into a tubular turbulent reactor), the mixing of the liquid flows is inefficient. Keeping the turbulent diffusion coefficient (D_T) constand along the reaction zone length provides sufficiently high effectivity of the mixing process. In this case, the rate of the main flow supply and the geometry and design of a tubular turbulent reactors play an essential part in obtaining high degrees of mixing.

Additional problems appear when it is necessary to mix the flows of liquids different in density and viscosity.

Still another problem is the possibility of controlling the heat removal in fast chemical processes, including polymerization. Fast processes and isobutylene polymerization in particular are uncontrollable in modern industries. Temperature variation in the reaction zone approaches 80–100°, there is local overheating of the reaction mixture, and the temperature fields from along the reactor coordinates. Fast and superfast polymerization processes proceeding in small-size tubular reactors in high turbulent flows, by contrast, become controlled and monitored as regards both the temperature and the concentration parameters of the process. Molecular-mass characteristics of the polymer products formed involving internal and external heat removal are also controlled, especially when the process is run under the conditions of the new macroscopic regime of ideal displacement in turbulent flow.

The convenient way to control the temperature in the reaction zone appears to be through solvent boiling, since the boiling temperature of a chemical compound depends on the pressure in the system. The chemical nature of the solvent and the pressure in the reactor make it possible to cover a rather wide range of temperatures. That is a sufficiently effective way to regulate temperatures in very fast chemical processes, including polymerization.

The external heat removal is, as a rule, less efficient. However, in realizing the regime of quasi-ideal displacement ($R < R_{cr}$) when the planar reaction front is formed and the temperature flow in the reaction zone is relatively uniform along the reactor zone radius, the process run, as well as the molecular mass characteristics of the polymer formed, becomes, independent of the initial temperature of raw materials within a certain temperature interval. The monomer polymerization proceeds under nearly isothermal conditions, with the polydispersion index ($\bar{P}_W/\bar{P}_n$) of the product approaching the most probable value but depending on the difference between the initial raw material temperature and the boiling temperature of the solvent and monomer.

The intensification of the mass exchange results in the marked increase of the efficiency of the external heat removal, which, in its turn, allows us to reduce the reaction temperature and, correspondingly, to increase the average MW with the simultaneous narrowing of the MWD of the polymer formed.

The possibility of carrying out fast polymerization process in nearly isothermal in conditions in a turbulent flow gives use to the problem of creating and developing new principles, high productivity, and efficient technology of the polymer production based on the process proceeding in a turbulent flow or turbulent vortex, that ensures the heat and mass transfer commensurable in time with the chemical reaction. This development can be carried out, in particular, in processes of isobutylene and butene polymerization.

The new technology of isobutylene and butene polymerization characterized by using small-size tubular reactors ensures the retention of basic principles of modern technological procedure, the reduction of the industrial premises and workers, the increase of the total productivity of processes five-fold or more, the possibility of using the wide-range composition raw monomer, including the non-standard fractions C_4-hydrocarbon one without the additional treatment, the universal character of the process, the reduction of the expence coefficients of the raw monomer and catalyst, the drop of the electro-power consumption, etc.

The universal character of the developed technology using tubular turbulent reactors has been confirmed by the emergenece of new procedures, methods and devices. This holds true not only in realizing very fast polymerization but also in conducting fast chemical and physical processes with low-molecular compounds, when the mass exchangs is the limiting stage.

A series of tubular turbulent reactors and apparatus have been introduced in the commercial production of oligo- and polyisobutylenes, low-molecular butyl rubber, stereoregular polyisoprene rubbers obtained on the "Al-Ti" and "Al-rare-earth elements" catalytic systems: of emulsified divinul-α-methyl styrene rubbers; ethyl chloride and dichloroethane by the reaction of the liquid-phase hydrochlorination and chlorination of ethylene, of metylethyl ketone at the stage of the production of butyl sulfuric acid as well as in the extraction of phenols and pyridines from the coal resin in the cokery.

The prospects of using small-size tubular turbulent reactors-mixers-extractors-heat exchangers in many branches of the modern industry seem be inexhaustible.

REFERENCES

[1] Kennedy, J. P., Cationic polymerization of olefins: a critical review, M: Mir, 1978, –430 p.
[2] Minsker, K. S., Sangalov, Ju. A., Isobutylene and its polymers., M: Khimia, 1986, –224 p.
[3] Frank-Kamenetskii, D. A., Diffusion and heat transfer in the chemical kinetics. –Second edition. –M: Nauka, 1967, –490 p.
[4] Zeldovitch, Ya. B., Barenblatt, G. I., Librovich, V. B., Makhviladze, G. M., Mathematical theory of burning and explosion. –M: Nauka, 1980, –478 p.
[5] Papisov, I. M., Kabanov, V. A., Kargin, V. A., Vysokomol. Soedin., 1965, v. 7, No. 10, pp. 1779–1785.
[6] Yenikolopyan, N. S., Khzardzhan, A. A., Gasparyan, E. E., Volyeva, V. B., Dokl. Akad Nauka SSSR, 1987, –v. 294, No. 5, pp. 1151–1154.
[7] Yerusalimskii, B. L., Lubetskii, S. G., Processes of the ionic polymerization. –L: Khimia, 1974, –256 p.
[8] Scwarc, F. R. S., Anionic polymerization. Carbanions, living polymers and electron transfer processes. –M: Mir, 1971. –669 p.
[9] Yenikolopyan, N. S., Volfson, S. A., Chemistry and technology of polyformaldehide. –M: Mir, 1968, –279 p.
[10] Goto, S., Yamomoto, K., Furui, S., Sugitomo M., J. Appl. Polym. Sci., Appl Polym Symp, 1981, v. 36, No. 1, pp. 21–40.
[11] Kabanov, V. A., Zubov, V. P., Semchikov Yu. D., Complex-radical polymerization., M: Khimia, 1987, –256 p.
[12] Korshak, V. V., Unbalanced polycondensation., M: Nauka, 1978, –696 p.
[13] Shimomura, T., Tolle, K. J., Smud, J., Szwarc, M., J. Am Chem Soc., 1967, v. 89, No. 4, pp. 796–803.
[14] Hostalka, H., Schulz, G. V., J. Polym. Sci., 1965, B. v. 3, No. 12, pp. 1043–1044.
[15] Shimomura, T., Smid, J., Szwarc, M., J. Am Chem Soc, 1967, v. 89, No. 23, pp. 5743–5749.

[16] Lee, C. L., Smid, J., Szwarc M., Trans Faraday Soc, 1963, v. 59, No. 5, pp. 1192–1200.
[17] Eizner, Yu. E., Yerusalimskii, B. L., Vysokomol. Soedin, 1970, A, v. 12, No. 7, pp. 1614–1620.
[18] Taylor, R. B., Williams, F., J. Am Chem Soc, 1969, v 91, No. 14, pp. 3728–3732.
[19] Kennedy, J. P., Shinkawa, A., Williams, F. J., Polym Sci, 1971, A-1, v. 9, No. 6, pp. 1551–1561.
[20] Prochukhan, Yu. A., Berlin, Al. Al., Minsker, K. S., Yenikolopyan, N. S., Dokl Akad Nauk SSSR, 1986, v. 287, No. 3, pp. 682–685.
[21] Berlin, Al. Al., Prochukhan, Yu. A., Minsker, K. S., Yenikolopyan, N. S., Internat. Polym Sci and Technology, 1986, v. 13, No. 12, pp. 90–93.
[22] Prokofyev, K. V., Verbitskii, B. G., Rogov, S. A., Kirichenko, L. N., Low-molecular polyisobutylenes. Topical review. M:, ZNIITENeftechim, 1982, –51 p.
[23] Minsker, K. S., Berlin, Al. Al., Svinukhov, A. G., Prochukhan, Yu. A., Yenikolopyan, N. S., Dokl Akad Nauk SSSR, 1986, v. 286, No. 5, pp. 1171–1173.
[24] Vetlinger, S. A., Permeability of polymer materials, M:, Khimia, 1974, –296 p.
[25] Berlin, Al. Al., Minsker, K. S., Sangalov, Yu. A., Novikov, D. D., Poznjak, T. I., Prochukhan Yu. A., Kirillov, A. P., Svinukhov, A. G., Vysokomol. Soedin, 1979, B. v. 21, No. 6, pp. 468–471.
[26] Poznjak, T. I., Lisitsyn, D. M., Novikov, D. D., Dyachkovskii, F. S., Vysokomol Soedin, 1977, A, v. 19, No. 5, pp. 1168–1170.
[27] Poznjak, T. I., Lisitsyn, D. M., Novikov, D. D., Berlin, Al. Al., Dyachkovskii, F. S., Prochukhan, Y. A., Sangalov, Yu. A., Minsker, K. S., Vysokomol. Soedin., 1980, A, v. 22, No. 6, pp. 1424–1427.
[28] Berlin Al. Al., Minsker K. S., Sangalov, Yu. A., Svinukhov, A. G., Kirillov, A. P., Yenikolopyan, N. S., Vysokomol Soedin, 1980, A, v. 22, No. 3, pp. 566–574.
[29] Prochukhan, Yu. A., Minsker, K. S., Berlin, Al. Al., Karpasas, M. M., Kompaniets, V. Z., Konoplev, A. A., Yenikolopyan, N. S., Dokl Akad Nauk SSSR, 1988, v. 298, No. 6, pp. 1428–1430.
[30] Zeldovich, Ya. B., Chossen of works: Chemical physics and hydrodynamics, M:, Nauka, 1984, –374p.
[31] Berlin, Al. Al., Minsker, K. S., Prochukhan, Yu. A., Karpasas, M. M., Yenikolopyan, N. S., Dokl Akad Nauk SSSR, 1986, v. 287, No. 1, pp. 145–148.
[32] Berlin, Al. Al., Minsker K. S., Prochkhan, Yu. A., Karpasas, M. M., Yenikolopyan N. S., Vysokomol Soedin, 1986, B, v. 28, No. 6, pp. 461–465.
[33] Berlin, Al. Al., Minsker, K. S., Prochukhan, Yu. A., Karpasas, M. M., Yenicolopyan, N. S., Internat Polym Sci and Technology, 1986, v. 13, No. 11, pp. 95–97.
[34] Minsker, K. S., Berlin, Al. Al., Prochukhan, Yu. A., Yenikolopyan, N. S., Vysokomol Soedin, 1986, B, v. 28, No. 6, pp. 466–469.
[35] Berlin, Al. Al., Prochukhan, Yu. A., Minsker, K. S., Yenikolopyan, N. S., Internat. Polym Sci and Technology, 1986, v. 13, No. 12, pp. 90–93.
[36] Kompaniets, V. Z., Konoplev, A. A, Berlin, Al. Al., Prochukhan, Yu. A., Minsker, K. S., Karpasas, M. M., Yenikolopyan N. S., Dokl Akad Nauk SSSR, 1987, v. 297, No. 5, pp. 1129–1132.
[37] Berlin, Al. Al., Kompaniets, V. Z., Konoplev, A. A., Minsker, K. S., Minsker, S. K., Prochukhan, Yu. A., Ryabenko, E. A., Yenikolopyan, N. S., Dokl Akad Nauk SSSR, 1989, v. 305, No. 5, pp. 1143–1146.
[38] Berlin, Al. Al., Prochukhan, Yu. A., Minsker, K. S., Karpasas, M. M., Bakhitova, R. Kh., Yenikolopyan, N. S., Vysokomol Soedin, 1988, A, v. 30, No. 6, pp. 1259–1262.
[39] Berlin, Al. Al., Prochukhan, Yu. A., Tumanyan, E. A., Minsker, K. S., Alexanyan, G. G., Yenikolopyan, N. S., Vysokomol Soedin, 1988, A, v. 30.
[40] Minsker K. S., Berlin, Al. Al., Prochukhan, Yu. A., Tumanyan, E. A., Karpasas, M. M., Yenikolopyan, N. S., Dokl Akad Nauk SSSR, 1986, v. 291, No. 1, pp. 114–119.
[41] Prochukhan, Yu. A., Minsker, K. S., Berlin, Al. Al., Tumanyan, E. A., Yenikolopyan, N. S., Dokl Akad Nauk SSSR, 1986, v. 291, No. 6, pp. 1425–1428.

[42] Berlin, Al. Al., Prochukhan, Yu. A., Minsker, K. S., Alexanyan, G. G., Grobov, S. V., Yenikolopyan N. S., Vysokomol Soedin, 1989, A, v. 31, No. 3, pp. 612–616.
[43] Kompaniets, V. Z., Konstantinov, A. A., Zymbalyuk, M. Ja., Epstein I. L., Chemical reactions in low temperature plasma, M:, Nauka, 1977, pp. 10–25.
[44] Kompaniets, V. Z., Konoplev, A. A., Polak, L. S., Potapova, E. V., Experimental and theoretical investigation of plasmochemical processes, M:, Nauka, 1984, pp. 18–38.
[45] Olevskii, V. M., Ruchinskii, V. P., Rotor-film heat- and mass-exchange apparatus, M:, Khimia, 1977, –200 p.
[46] Budtov, V. P., Kansetov, V. V., Heat-and mass-transfer in polymerization processes, L:, Khimia, 1988, –256 p.
[47] Fedorov, A. Ya., Litvak, G. E., Vysokomol Soedin, 1991, A, v. 33, No. 12, pp. 2626–2634.
[48] Protod'yakonov, I. O., Ul'yanov, S. V., Hydrodynamics and mass-exchange in "liquid–liquid" dispersed sistems, L:, Nauka, 1987, –348 p.
[49] Midlman, S., Flowing of polymers, M:, Mir, 1971, –260 p.
[50] Abramovich, G. N., Theory of turbulent flows, M:, Physmatgiz, 1960, –705 p.
[51] Barchilon, M., Kurge, P., Theoretical base of engineering calculations, 1964, No. 4, pp. 173–184.
[52] Shez, J., Turbulent flow, Processes of blowing and mixing, M:, Mir, 1984, –247 p.
[53] Thermodynamic properties of individual compounds (Reference book in 4 volumes)/Ed. Glushko, V. P., M:, Nauka, 1979, v. 2, book 2, –344 p.
[54] Berlin, Al. Al., Minsker, K. S., Prochukhan, Yu. A., Yenikolopyan, N. S., Vysokomol Soed, 1989, A, v. 31., No. 9, pp. 1779–1798.
[55] Minsker, S. K., Golubeva, T. V., Konopljov, A. A., Kompaniets, V. Z., Berlin, Al. Al., Minsker K. S., Yenikolopyan N. S., Dokl Akad Nauk SSSR, 1990, v. 314, No. 6, pp. 1450–1454.
[56] Minsker S. K., Konopljov, A. A., Minsker K. S., Prochukhan, Yu. A., Kompaniets, V. Z., Berlin, Al. Al., Theoretical basis of chemical technology, 1992, v. 26, No. 5, pp. 686–691.
[57] Minsker, S. K., Konopljov, A. A., Golubeva, T. V., Prochukhan, Yu. A., Kompaniets, V. Z., Zaikov, G. E., Berlin, Al. Al., Polymer Yearbook-10, Ed. Pethrick, R. A., Switzerland et al., Harwood Acad Publ, 1993, pp. 93–100.
[58] Berlin, Al. Al., Volfson, S. A., Kinetic methods in synthesis of polymers, M:, Khimia, 1973, –344 p.
[59] Minsker, K. S., Berlin, Al. Al., Prochukhan, Yu. A., Karpasas, M. M., Yenikolopyan, N. S., Vysokomol. Soedin., 1986, A, v. 28, No. 6, pp. 461–465.
[60] Prochukhan, Yu. A., Minsker, K. S., Berlin, Al. Al., Yenikolopyan, N. S., Dokl. Akad. Nauk SSSR, 1988, v. 298, No. 6, pp. 1428–1430.
[61] Berlin, Al. Al., Karpasas, M. M., Kompaniets, V. Z., Konopljev, A. A., Minsker, K. S., Prochukhan, Yu. A., Yenikolopyan, N. S., Dokl Akad Nauk SSSR, 1989, v. 305, No. 2, pp. 365–368.
[62] Berlin, Al. Al., Prochukhan Yu. A., Minsker, K. S., Alexanyan, G. G., Grobov, S. V., Yenikolopyan, N. S., Vysokomol Soedin, 1988, A, v. 30, No. 11, pp. 2441–2446.
[63] Guterbok, G., Polyisobutylene and copolymers of isobutylene, L:, Gostoptekhisdat, 1962, –363 p.
[64] Minsker, K. S., Prochukhan, Yu. A., The creation of unified energy- and material-saving technologies of new generation for very fast chemical processes and their realization at chemical industries of the Republic of Bashkortostan: Materials of I congress of chemists, petrochemists and oil refinery engineers Republic Bashkortostan, Ufa, 1992, pp. 45–53.
[65] Berlin, Al. Al., Minsker, K. S., Prochukhan, Yu. A., Karpasas, M. M., Yenikolopyan, N. S., Internat. Polym Sci and Technology, 1986, v. 13, No. 11, pp. 95–97.
[66] Berlin, Al. Al., Minsker, K. S., Wirtschaft Technik, 1990, No. 2, pp. 54–56.
[67] J. of national Mendeleevs Chem Soc, 1989, v. 34, No. 6, pp. 71.
[68] Berlin, Al. Al., Prokof'ev, K. V., Minsker, K. S., Prochukhan, Yu. A., Kotov, S. V.,

Bulankov, V. F., Yasinenko, V. A., Naumova, T. I., Chemistry and technology of oils and fuels, 1988, No. 7, pp. 8–9.
[69] Berlin, Al. Al., Minsker, K. S., Prokof'ev, K. V., Prochukhan, Yu. A., Sangalov, Yu. A., Petroleum refining and petrochemisty, 1988, No. 2, pp. 25–28.
[70] Kotov, S. V., Prokof'ev, K. V., Prochukhan, Yu. A., Minsker, K. S., Adilov, N. A., Velishov, D. A., Petroleum refining and petrochemistry, 1989, No. 1, pp. 17–18.
[71] Minsker, K. S., Prochukhan, Yu. A., Kolesov, S. V., Berlin, Al. Al., Kutlugugina, I. Kh., Prokof'ev, K. V., Kozlov, V. G., Yesinenko, V. G., Alexanyan, G. G., Petroleum refining and petrochemistry, 1989, No. 11, pp. 40–42.
[72] Prokof'ev, K. V., Kotov, S. V., Berlin, Al. Al., Minsker, K. S., Prochukhan, Yu. A., Sangalov, Yu. A., Bull. MChT Interproduct, Sofia, Bulgaria, 1989, No. 1, pp. 19–22.
[73] Kotov, S. V., Prokof'ev, K. V., Minsker, K. S., Sangalov, Yu. A., Berlin, Al. Al., Chemistry and technology of oil and fuels, 1990, No. 4, pp. 14–15.
[74] Kotov, S. V., Berlin, Al. Al., Prokof'ev, K. V., Minsker, K. S., Sangalov, Yu. A., Prochukhan, Yu. A., Adilov, N. A., Alexanyan, G. G., Yasinenko, V. A., Chemistry and technology of oil and fuels, 1990, No. 6, pp. 10–11.
[75] Kotov, S. V., Atmatgev, B. E., Minsker, K. S., Yasinenko, V. A., Prokog'ev, K. V., Berlin, Al. Al., Petroleum refining and petrochemistry, 1992, No. 1, pp. 38–41.
[76] Berlin, Al. Al., Prochukhan, Yu. A., Minsker, K. S., Yenilkolopyan, N. S., Vysokomol. Soedin., 1991, A, v. 33, No. 2, pp. 243–270.
[77] Berlin, Al. Al., Minsker, K. S., Prochukhan, Yu. A., Yenikolopyan, N. S., Polym.–Plastics Technol and Eng, 1991, v. 30, No. 2/3, pp. 253–297.
[78] Kompaniets, V. Z., Ovsyanikov, A. A., Polak, L. S., Chemical reactions in turbulent flow of gas and plasma, M:, Nauka, 1979, –242 p.

INDEX